Kissin

Polyethylene

Yury V. Kissin

Polyethylene

End-Use Properties and their Physical Meaning

2nd Edition

HANSER

Hanser Publishers, Munich

Hanser Publications, Cincinnati

The Author:

Yury V. Kissin,
Rutgers, The State University of New Jersey, Department of Chemistry and Chemical Biology,
Piscataway, New Jersey 08854-8087, USA

Distributed in the Americas by:
Hanser Publications
414 Walnut Street, Cincinnati, OH 45202 USA
Phone: (800) 950-8977
www.hanserpublications.com

Distributed in all other countries by:
Carl Hanser Verlag
Postfach 86 04 20, 81631 Munich, Germany
Fax: +49 (89) 98 48 09
www.hanser-fachbuch.de

Library of Congress Control Number: 2020942243

Editor: Dr. Mark Smith
Production Management: Der Buchmacher, Arthur Lenner, Windach
Coverconcept: Marc Müller-Bremer, www.rebranding.de, Munich
Coverdesign: Max Kostopoulos
Typesetting: Kösel Media, Krugzell, Germany
Printed and bound by Druckerei Hubert & Co GmbH und Co KG BuchPartner, Göttingen
Printed in Germany

ISBN: 978-1-56990-831-0
E-Book ISBN: 978-1-56990-832-7

To Samuel and Gabriel

Preface

Every commercial information sheet describing a polyethylene resin, every professional discussion of the application range for a resin, every issue of the resin pricing – they all involve the same short list of resins' end-use properties. Although the items in the list may vary depending on application, these end-use properties are universally understood throughout the "insider" word without any need for an explanation.

In the case of commercial film-grade LLDPE resins, these parameters usually include the resin's density, the melt index, the melt flow ratio, the melting point, the dart impact strength, the tear strength in two directions of the film, film haze, and extractables. In the case of blow-molding HDPE resins, the standard parameters are the density, the high-load melt index, environmental stress cracking resistance, the top-load strength, and so on. Tables P.1, P.2, and P.3 show three representative examples of these parameters taken from commercial product data-sheets.

These engineering terms and these values have become "the lingua franca" of all product engineers, plant operators, and catalyst chemists throughout the world involved in the production and testing of polyethylene resins. Test procedures for these parameters are minutely described in standardized documents, ASTM methods in the US and ISO methods in Europe.

On the other hand, the polymer science has developed a different universally understood set of parameters that characterize the same polymers: the average molecular weight, the molecular weight distribution, the content of an α-olefin in an ethylene/α-olefin copolymer, compositional uniformity of the copolymers, stress-strain characteristics of semi-crystalline polymers, etc.

Table P.1 ExxonMobil LLDPE Resin, LL 1001 Series (Ziegler-Natta Catalyst)

	SI units	English units	Measurement method
Density	0.918 g/cm^3	0.0332 lb/in^3	ASTM D1505
Melt index (190 °C, 2.16 kg)	1.0 g/10 min	1.0 g/10 min	ASTM D1238
Peak melting temperature	121 °C	250 °F	
Tensile strength at yield, MD	9.4 MPa	1,400 psi	ASTM D882
Tensile strength at yield, TD	9.5 MPa	1,400 psi	ASTM D882
Tensile strength at break, MD	50 MPa	7,700 psi	ASTM D882
Tensile strength at break, TD	35 MPa	5,100 psi	ASTM D882
Elongation at break, MD	580%	580%	ASTM D882
Elongation at break, TD	850%	850%	ASTM D882
Secant modulus, MD, at 1%	190 MPa	28,000 psi	ASTM D882
Secant modulus, TD, at 1%	220 MPa	32,000 psi	ASTM D882
Dart drop impact strength	100 g	100 g	ASTM D1709A
Elmendorf tear strength, MD	80 g	80 g	ASTM D1922
Elmendorf tear strength, TD	400 g	400 g	ASTM D1922
Haze	14%	14%	ASTM D1003

Table P.2 ExxonMobil LLDPE Resin Exceed 1018KA Series (Metallocene Catalyst)

	SI units	English units	Measurement method
Density	0.918 g/cm^3	0.0332 lb/in^3	
Melt index (190 °C, 2.16 kg)	1.0 g/10 min	1.0 g/10 min	ASTM D1238
Peak melting temperature	<119 °C	<247 °F	
Tensile strength at yield, MD	9.65 MPa	1,400 psi	ASTM D882
Tensile strength at yield, TD	9.65 MPa	1,400 psi	ASTM D882
Tensile strength at break, MD	54.5 MPa	7,900 psi	ASTM D882
Tensile strength at break, TD	42.7 MPa	6,200 psi	ASTM D882
Elongation at break, MD	500%	600%	ASTM D882
Elongation at break, TD	600%	600%	ASTM D882
Secant modulus, MD, at 1%	186 MPa	27,000 psi	ASTM D882
Secant modulus, TD, at 1%	193 MPa	28,000 psi	ASTM D882
Dart drop impact strength	590 g	1.3 lb	ASTM D1709A
Elmendorf tear strength, MD	250 g	250 g	ASTM D1922
Elmendorf tear strength, TD	470 g	470 g	ASTM D1922
Haze	18%	18%	ASTM D1003

Table P.3 Braskem Blow Molding HDPE Resin GM7746C

	SI units	English units
Density	0.944 g/cm³	0.0341 lb/in³
Melt index (190 °C, 21.6 kg)	4.5 g/10 min	4.5 g/10 min
Vicat softening point	126 °C	259 °F
Tensile strength at yield	23 MPa	3,340 psi
Tensile strength at break	42 MPa	6,090 psi
Elongation at yield	13%	13%
Elongation at break	880%	880%
Flexural (secant) modulus at 1%	890 MPa	129,000 psi
Secant modulus, TD, at 1%	220 MPa	32,000 psi
Environmental stress-cracking resistance (100% Igepal)	≥ 1,000 h	≥ 1,000 h

This book provides a necessary bridge between the meaning of the engineering end-use parameters of polyethylene resins and molecular and structural characteristics of the same materials used in the polymer science. The main goal is to translate such commonly used parameters as the melt index of a resin or the dart impact strength of a film sample into the universal language of the polymer science. After this translation is completed, many facets of the resin engineering properties became transparent and easily explainable. For example:

- What happens with the melt flow ratio of a resin after the catalyst used to produce it is modified to increase its sensitivity to an α-olefin?
- What happens with the dart impact strength or the tear strength of LLDPE film when butene is replaced with hexene or octene in an ethylene/α-olefin copolymerization reaction employing the same catalyst and why does it happen?
- Why are the melting points of metallocene LLDPE resins so much lower compared to the melting points of LLDPE resins of the same density and melt index prepared with Ziegler-Natta catalysts?
- Why are hexane extractables of metallocene LLDPE resins lower compared to extractables of LLDPE resins of the same density produced with Ziegler-Natta catalysts?

These are the types of questions this book provides answers to. Detailed analysis of many such links between end-use engineering properties of polyethylene resins and molecular characteristics of the polymers turn out to be quite complex. For this reason, the description of each such linkage is accompanied by numerous examples of practical significance and by explicit data for commodity polyethylene resins.

This book is written with three audiences in mind. The first, the most populous audience, includes product engineers, the specialists who evaluate properties of polyethylene resins and judge their usefulness (as well as pricing) for a particular application. These specialists are very adept at measuring and evaluating end-use engineering properties of the resins they are working with. However, they are usually less confident when asked which of the molecular characteristics of the polymers should be changed, and in what direction, to improve a particular end-use property of a resin.

The members of the second audience are plant and pilot plant operators in the polyethylene industry. These individuals deal with large-scale continuous production processes and need to know which of the process variables they control are crucial for achieving and maintaining the desired end-use parameters of the resins.

The members of the third audience are catalyst chemists, specialists in designing new polymerization catalysts and modifying the existing ones. These professionals often judge a success or a failure of a catalyst they develop based on properties of a small amount of polymer prepared in the laboratory, from ~ 10 to ~ 200 g. Their principal interest is to know which parameters of a polymer prepared in bench-type tests have the highest predictive power and how to translate the changes they make in a catalyst recipe into the changes in the end-use properties of the resins manufactured with the catalyst on the commercial scale. One has to take into account that measurements of some end-use properties require large quantities of resins far exceeding what can be prepared in laboratory.

This book is intended to improve communication bridges between these three groups of specialists and to aid them in understanding each other better and faster.

The second edition of the book contains updated information on many aspects of the relationships between scientific and engineering characterization of various polyethylene resins, as well as new information on polyethylene recycling (a rapidly expanding field of resin applications with a different set of technical requirements) and on polyethylene resins prepared with post-metallocene catalysts. It also contains a new chapter describing several important commercial characteristics of polyethylene film, the content of extractable and soluble material in LLDPE film, blocking of LLDPE film, and film haze.

Yury V. Kissin

Rutgers, The State University of New Jersey
New Jersey, USA

Contents

1 Educational Minimum: Synthesis, Structure, and Properties of Polyethylene Resins

1.1 Classification and Applications of Polyethylene Resins

The term "polyethylene resins" describes catalytically produced semi-crystalline homopolymers and copolymers derived mostly from ethylene [1–4], as well as ethylene polymers produced in radical polymerization reactions under high pressure [4–6]. Polyethylene resins of different types are widely used as commodity plastics.

Some polyethylene resins contain strictly linear polymer chains; their chemical formula is $-(CH_2-CH_2)_n-$, where n is a very large number, from ~ 1,000 to ~ 100,000. Other polyethylene resins contain branches in their chains. Most such resins are produced in ethylene/α-olefin copolymerization reactions. The molecular structure of ethylene/α-olefin copolymers with a low α-olefin content can be approximately represented by the formula:

$$\sim-(CH_2-CH_2)_x-CH_2-CH(Branch)-(CH_2-CH_2)_y-CH_2-CH(Branch)-(CH_2-CH_2)_z-CH_2-CH(Branch)-\sim$$

where the $-CH_2-CH_2-$ units come from ethylene and the $-CH_2-CH(Branch)-$ units come from the α-olefin molecule. The x, y, and z values in these chains can vary from small (4 to 5) to a very large number.

All the branches in catalytically produced polyethylene resins are of the same size; they are the alkyl substituents in the α-olefin molecules: the ethyl group if the α-olefin is 1-butene, the butyl group if the α-olefin is 1-hexene, the hexyl group if the α-olefin is 1-octene, or the isobutyl group if the α-olefin is 4-methyl-1-pentene. When ethylene is polymerized at a high pressure via the radical mechanism, they have essentially the same chain structure. In this case, branches of several different types are formed spontaneously due to peculiarities of the radical polymerization reactions. These branches are linear or branched alkyl groups [7, 8]. Their lengths vary widely within each polymer molecule. Two types of such branches are distinguished: short-chain branches, from the methyl to the isooctyl group, and long-chain branches, up to several thousand carbon atoms in length. The most abundant among the short-chain branches are butyl branches (~ 5 to 6 per 1,000C), ethyl branches (~ 4 per 1,000C), and pentyl branches (~ 2 per 1,000C), whereas the content of hexyl and longer linear branches is lower, 1 to 2 per 1,000C.

The level of branching in catalytically produced polyethylene resins is defined differently in industry and in polymer science. In polymer science, these resins are called ethylene/α-olefin copolymers with a particular molar content of α-olefin, C_M^{copol}, mol %. In industry, the level of branching in polyethylene resins is usually represented by the value called the *branching degree*. It is defined as the number of branches per 1,000 carbon atoms, Branch/1,000C. Because all such branches end with the methyl group, the branching degree is often represented by the symbol CH_3/1,000C. The ratio between C_M^{copol} and the branching degree is given by:

$$CH_3/1{,}000C = C_M^{copol} / (0.2 + 0.001 \cdot k \cdot C_M^{copol}) \qquad (1.1)$$

The k parameter in Equation 1.1 is equal to 2 for ethylene/1-butene copolymers, 4 for ethylene/1-hexene copolymers, and 6 for ethylene/1-octene copolymers. Polyethylene resins produced in radical reactions at a high pressure have branches of different size and their branching level is universally expressed as the branching degree, CH_3/1,000C.

Commercial classification of polyethylene resins has developed historically, in parallel with the discovery of new catalysts for ethylene polymerization and new polymerization processes. The classification is based on two parameters that could be easily measured in the 1950s in a commercial environment with a minimum of instrumentation: the rheological parameter called the *melt index*, which reflects the average molecular weight of the resin, and the *resin density* (the function of its crystallinity degree). The physical meaning of these two seemingly straightforward parameters is in reality quite complex. They are described in Chapters 2 and 4, respectively.

According to the American Society of Testing and Materials (ASTM D1248-05 and ASTM-D3350), all polyethylene materials are divided into various classes. The classification specifies resins with uniform sets of properties. Density determines the type of the resin and the melt index determines its category. The commonly used commercial classification is given in Table 1.1.

HDPE resins with nominal density of 0.941 to 0.959 g/cm^3 belong to Type III and those with nominal density higher than 0.960 belong to Type IV. Some high molecular weight resins with density lower than 0.941 g/cm^3 and the resins with an ultrahigh molecular weight and with density of ~ 0.930 g/cm^3 are also represented as HDPE resins because of their low branching degree, although they formally belong to Type II resins.

Table 1.1 Commercial Classification of Polyethylene Resins

Resin type	Symbol	α-Olefin content, mol %	Crystallinity degree, %	Density, g/cm^3
Resins of high density	HDPE	0	65 to 70	>0.960
Resins of high density	HDPE	0.2 to 0.5	60 to 65	0.941 to 0.959
Resins of ultrahigh molecular weight	UHMW HDPE	0	30 to 40	0.930 to 0.935
Resins of medium density	MDPE	1 to 2	45 to 55	0.926 to 0.940
Resins of low density	LLDPE	2.5 to 3.5	30 to 45	0.915 to 0.925
Resins of very low density	VLDPE	>4	<25	<0.915
Low density polyethylene produced in radical reactions	LDPE	Branching degree 20 to 30 CH_3/1,000C	45 to 55	0.910 to 0.940

The content of an α-olefin in commercially manufactured ethylene/α-olefin copolymers varies in a wide range, from <0.5 up to 20 mol %. These copolymers, depending on the content of α-olefin, are called *medium density polyethylene resins* (MDPE), *linear low density polyethylene resins* (LLDPE), or *very low density polyethylene resins* (VLDPE). The group of the VLDPE resins is further divided into two subgroups, polyethylene *plastomers* with a crystallinity degree of 10 to 20% and density from 0.915 to 0.900 g/cm^3, and completely amorphous ethylene *elastomers* with density as low as 0.86 g/cm^3. By definition, all the catalytically produced resins contain only short-chain branches derived from α-olefins. However, polymerization reactions utilizing some metallocene catalysts and chromium oxide catalysts can also introduce long-chain branches in the polyethylene chains (Section 1.4).

Some metallocene catalysts can copolymerize ethylene with cycloolefins, such as cyclopentene, cyclooctene, or norbornene. In this case, the branches in polyethylene chains are either small cycles containing from 5 to 10 carbon atoms, or two fused cycles. These materials form an additional resin type called *cycloolefin copolymers* (COC).

The five categories of polyethylene resins are specified according to their melt index measured according to ASTM D1238-10:

Category:	1	2	3	4	5
Melt index, g/10 min:	>25	10 to 25	1 to 10	0.4 to 1.0	below 0.4

Other characteristics of polyethylene resins, predominantly color, are specified by class. The three classes of polyethylene resins are designated as A, B, and C. The classes indicate color, amounts, and types of antioxidants, and other additives.

Class A refers to naturally colored polyethylene resins, Class B includes white and black-colored resins, and Class C covers weather-resistant black resins containing more than 2% carbon black.

The classification of polyethylene resins in its present form affords a basic distinction between different resin types. However, the classification is often poorly suited to delineate fine differences between structures and properties of various resins that play an important role in the sophisticated modern resin market. After all, the market grades different resins mostly according to their end-use properties rather than by their general classification.

Taken together, polyethylene resins account for the largest fraction, ~35%, of the worldwide plastic production. The volume of HDPE resins manufactured worldwide in 2019 was over 50 million metric tons and the combined volume of LLDPE and LDPE resins over 60 million metric tons. In the US alone, the total production volume of all polyethylene grades in 2019 amounted to ~23 million metric tons.

Applications of polyethylene resins vary greatly by the grade. The applications of the two most important grades are:

HDPE resins	Blow molding (containers and bottles)	~31%
	Film (biaxially oriented)	~28%
	Injection molding	~22%
	Other applications	~19%
LLDPE resins	Film (blown and cast)	~80%
	Injection molding	~7%
	Wire and cable coating/insulation	~4%
	Other applications	~9%

■ 1.2 Catalysts for Synthesis of Polyethylene Resins

Polyethylene resins are produced commercially with transition metal catalysts of different types [1–4, 9, 10] or in high-pressure radical polymerization reactions [4, 5]. This section provides merely a brief sketch of the catalyst chemistry; a much more detailed discussion of catalyst preparation techniques is presented in reviews [3, 4, 9, 10].

Several groups of polymerization catalysts are especially important.

Titanium-based Ziegler-Natta catalysts: All these catalysts consist of two components. The first component is called a *catalyst*, it is a liquid or a solid powder that contains a derivative of titanium [4, 9, 10]. The second component is called a *cocatalyst.* The cocatalyst is an organoaluminum compound, usually triethylaluminum. The preferred physical form of the catalyst depends on the type of the polymerization process it is used in (Section 1.3). Most modern Ziegler-Natta catalysts are supported and contain titanium tetrachloride as the active ingredient. A variety of techniques were developed for supporting titanium compounds on porous solids with a high internal surface area, such as silica and microcrystalline magnesium dichloride [1, 3, 10]. Ziegler-Natta catalysts are used to synthesize several grades of HDPE and LLDPE resins. In industry, polyethylene resins are produced with these catalysts at temperatures between 80 and 95 °C and at an ethylene pressure of 0.7 to 1.5 MPa (100 to 200 psi). Ethylene polymers synthesized under these conditions intrinsically have a very high molecular weight (see Section 1.4) and in order to decrease it, polymerization reactions with Ziegler-Natta catalysts are nearly always carried out in the presence of hydrogen, an effective molecular weight-controlling agent. In addition to polyethylene resins, Ziegler-Natta catalysts are also used for the commercial production of other important polyolefins including isotactic polypropylene, polybutene, and polymethylpentene, as well as synthetic rubbers based on polybutadiene and polyisoprene.

A different type of titanium-based Ziegler-Natta catalysts is used in industry in solution polymerization processes. These polymerization reactions are performed in heavy hydrocarbons (mixtures of C_8 to C_{10} alkanes) at high temperatures of 130 to 200 °C, and at a reactor pressure of 3.5 to 20 MPa (500 to 3,000 psi). Because titanium-based active centers are not stable at such high temperatures, catalyst formulations of very high activity are employed in these processes. The catalysts are formed directly in the polymerization reactors by combining three hydrocarbon-soluble components: an organomagnesium compound, such as dibutyl magnesium, a source of chlorine atoms, such as butyl chloride (these two compounds rapidly react to produce a catalyst support of finely dispersed magnesium dichloride), and the active ingredient, usually titanium tetrachloride.

Chromium oxide catalysts (Phillips catalysts): Chromium oxide catalysts are used for the synthesis of HDPE resins of various grades, particularly for the manufacture of ethylene homopolymers with a broad molecular weight distribution. All chromium oxide catalysts are supported; they use inert porous substrates, usually silica, and chromium compounds as active ingredients. The synthesis procedure of chromium oxide catalysts consists of two main steps. First, the support is impregnated with a source of chromium, such as water solution of chromic acid or chromium nitrate. Then the solvent is removed and the catalyst is activated by calcination at 500 to 850 °C in an oxidizing environment. Catalysts of a similar composition are also used for the synthesis of ethylene/α-olefin copolymers. These copolymers are usually designated as *low density linear polyethylene* (LDLPE) resins to distinguish them from LLDPE resins produced with Ziegler-Natta catalysts.

Organochromium catalysts: Several commercially important catalysts for the synthesis of injection molding HDPE resins utilize organochromium compounds, such as bis(triphenylsilyl)chromate or bis(cyclopentadienyl)chromium. All these catalysts are silica-supported.

Metallocene catalysts: Metallocene catalysts are used for the synthesis of HDPE and LLDPE resins of high compositional uniformity and a narrow molecular weight distribution. These catalysts are employed either in the soluble form or they are supported on inert carriers such as silica. The catalysts contain two components. The first component is a metallocene complex of zirconium or titanium. The second component, a *cocatalyst*, is either a special organoaluminum compound, methyl alumoxane (abbreviated as MAO), or a fluoroboronaromatic compound. Metallocene catalysts are also used to produce crystalline syndiotactic polypropylene and polystyrene, ethylene/propylene elastomers, and engineering ethylene/cycloolefin (COC) plastics.

Bimetallic catalysts: Several HDPE applications require polyethylene resins with an especially broad, preferably bimodal molecular weight distribution. Such materials cannot be produced simply by physical mixing of pellets of two HDPE resins with vastly different molecular weights. Instead, these resins are either obtained using Ziegler-Natta catalysts in special technological processes employing two or several polymerization reactors in a series or in a single reactor using supported bimetallic catalysts. The idea behind such catalysts is to combine within each catalyst particle active ingredients of two different types that operate under identical conditions but produce polymer material with different molecular weight.

Post-metallocene catalysts: The term "post-metallocene catalysts" has a historical rather than any particular chemical meaning, simply because these catalysts were developed several decades after the discovery of metallocene catalysts. Post-metallocene catalysts also consist of two components. The first component is a transition metal complex containing a bidentate or a multidentate organic ligand and the second component is an organoaluminum activator such as MAO or an ion-forming compound. The transition metal atoms in the complexes vary from scandium to lanthanoid metals (mostly Ti, Zr, Ni, Pd, Co, Fe), and the ligands coordinate to the metal atoms through oxygen, nitrogen, phosphorus, or sulfur atoms [4, 11, 12]. The list of the ligands includes diphenoxy, diamine, phenoxy-imide, saliginine, α-diimine, and bis(imino)pyridyl ligands with a variety of alkyl and aryl substituents. All the complexes are soluble in aromatic solvents, and some are soluble in aliphatic hydrocarbons. Some post-metallocene catalysts already found industrial application [13–15].

Radical initiators: Several types of initiators are used in high-pressure radical polymerization reactions of ethylene [4–6]. Most of them are organic peroxides of the R–O–O–R′ type, chemical compounds containing the peroxide bond between two organic moieties. The peroxide bond readily dissociates at increased temperatures with the formation of two radicals, R-O• and R′-O•. Different organic peroxides decompose at a different rate depending on the type of the R and R′ groups. The choice of peroxide is determined by the design of a polymerization reactor and the desired properties (density, degree of branching, average molecular weight) of the LDPE resin.

1.3 Industrial Processes for the Manufacture of Polyethylene Resins

Several technological processes are used to manufacture polyethylene resins in catalytic and radical polymerization reactions [4, 10].

Polymerization reactions in slurry: Polymerization processes in hydrocarbon slurry, usually in light hydrocarbons such as isobutane, hexane, or heptane, were historically the first commercial ethylene polymerization processes. These processes use chromium oxide and Ziegler-Natta catalysts; they still enjoy high popularity due to their versatility and convenience of operation [2, 10, 16]. The slurry technology accounts for the manufacture of nearly 60% of all polyethylene resins. The slurry reactions afford the production of the full range of polyethylene resins, from low molecular weight waxes to resins with a very high molecular weight. These processes are also used to produce resins of ultrahigh molecular weight.

Several types of slurry polymerization processes are used in industry [10, 17, 18]:

1. Two types of loop reactors in which a diluent (a low-boiling or a high-boiling hydrocarbon) circulates at a high speed through a long circular pipe.
2. Continuous stirred-tank reactors with a high-boiling diluent.
3. "Liquid pool" processes in which the polymerization reaction takes place in a reactor filled with a light diluent such as propane or isobutane.

All these reactors usually operate at 80 to 90 °C and a total pressure of 1 to 3 MPa (150 to 300 psi). The residence time of catalyst particles in the reactors varies from 2 to 3 hours.

Polymerization reactions in the gas phase: Many polymerization catalysts, including both Ziegler-Natta catalysts, chromium oxide catalysts, and supported metallocene catalysts, were adopted for use in gas phase reactors [2, 4, 10, 19–25]. Gas-phase processes account for over 20% of the world polyethylene capacity. These processes are very economical due to the absence of the diluent recovery stage. They are flexible and can accommodate a large variety of supported catalysts capable of polymerization at a relatively low pressure.

A typical gas-phase reactor is a tall cylindrical tower with a diameter that ranges from 3 to 6 m (10 to 20 ft) and a height-to-diameter ratio of 6 to 7. The tower is half-filled with a bed of polymer particles 0.5–5 mm in size. The particles are agitated either by employing the fluidized-bed technique (blowing a stream of ethylene through numerous openings in the reactor's bottom) or with a mechanical stirrer. A fresh catalyst in the form of small spherical particles 200 to 100 micron in diameter is continuously added to the polymer bed, and polymer particles are continuously removed from the reactor. During their stay in the reactor (the average residence time is usually 1 to 4 hours), the catalyst particles circulate through the polymer bed and gradually increase in size as the polymerization reaction proceeds, but they retain their shape. The fluidized-bed reactors exhibit high versatility with respect to the type of polymerization catalyst they can accommodate.

Two-reactor polymerization processes: Several modifications of the gas-phase reactor technology were developed. In one modification, two gas-phase fluidized bed reactors are connected in a series [10, 26, 27]. A similar reactor scheme uses a combination of two different reactor types: a small slurry loop pre-reactor followed by one or two gas-phase fluidized-bed reactors [28, 29]. Both two-reactor processes use titanium-based Ziegler-Natta catalysts for the manufacture of a special grade of polyethylene materials called *bimodal polyethylene resins*. These resins are ~1:1 blends of two finely intermixed polymer fractions, one fraction with a very high molecular weight and another with a very low molecular weight. The fraction with a very low molecular weight (~10,000 to 20,000) is produced in one of the reactors in the presence of a large amount of hydrogen, and the second fraction with a very high molecular weight (~400,000 to 600,000) is produced in another reactor at a very low hydrogen pressure. The telltale signature of these polymers is the shape of their molecular weight distribution curves, which consist of two clearly identifiable components (Section 1.6.4).

Polymerization reactions in solution: Two solution polymerization technologies are practiced. Processes of the first type employ high molecular weight organic solvents, while processes of the second type use molten polyethylene itself as the polymerization medium [4, 10, 18, 30, 31].

Many hydrocarbons dissolve polyethylene at 120 to 150 °C. Because the viscosity of the polyethylene solution rapidly increases with molecular weight, solution polymerization processes are employed primarily for the production of low molecular weight resins. A variety of Ziegler-Natta and metallocene catalysts survive high temperatures for short periods of time and can be used in the solution processes [2, 10]. Reactions in solution are carried out at 150 to 200 °C at a total reactor pressure of ~5 to 10 MPa (720 to 1,500 psi); they are very fast with the residence time from 5 to 15 minutes. These processes require the removal of the solvent from the polymer as the last step.

High-pressure radical polymerization reactions: LDPE resins are produced in radical polymerization reactions at a high pressure. Two types of reactors are used: long tubular reactors with diameters of 2.5 to 6.5 cm (1 to 2.5 in) and lengths from 500 to 1,000 m (~6,500 to 13,000 ft) and stirred autoclaves with a volume of ~0.5 m^3. In both cases, the reactor pressure is very high, from 120 to 300 MPa (18,000 to 45,000 psi), and the temperature ranges from 130 to 350 °C, depending on the resin grade. Ethylene exists in the supercritical state under these conditions and it readily dissolves polyethylene. These polymerization reactions are initiated with oxygen or with organic peroxides. The reactions are very fast; the typical residence time varies from 2 to 10 minutes, which allows the production of up to 20 ton/hour of polyethylene in a single reactor.

1.4 Chemistry of Ethylene Polymerization Reactions

Polymerization reactions of ethylene and copolymerization reactions of ethylene and α-olefins with all classes of catalysts discussed in Section 1.2 are catalytic in the same sense as other catalytic reactions: a single active center in a catalyst produces many polymer molecules, one after another, over a period of time dictated by the reactor technology, from several minutes in high-temperature solution processes to several hours in gas-phase and slurry processes. The real chemical structure of active centers in polymerization catalysts is firmly established only for metallocene catalysts. A metallocene complex Cp_2ZrCl_2 (here the *Cp* symbol stands for any ligand with the cyclopentadienyl group) reacts with a cocatalyst, MAO, or with trimethylaluminum present in MAO, and is converted into an initial active center, a metallocenium cation $Cp_2Zr^+{-}CH_3$. The negatively charge in the center is positioned within the MAO molecule.

The structure of initial active centers in Ziegler-Natta and chromium polymerization catalysts is not yet known definitely but it is generally accepted that they also contain the M–C bond and can be generally represented by the formula $[M]^+{-}R$. Here R is a small alkyl group (the starting end-group of a polymer chain, for example, the CH_3 group in the metallocene catalyst), M is a Ti or a Cr atom, and the brackets [ ] signify that other atoms attached to the M atom, as well the nature of the negatively charged counter-ion, are not yet definitely known.

The principal chemical step in any catalytic polymerization reaction of ethylene is the insertion of the double bond of an ethylene molecule into the $[M]^+{-}C$ bond:

$$[M]^+{-}\mathbf{C} + CH_2{=}CH_2 \; -(k_p)\rightarrow \; [M]^+{-}CH_2{-}CH_2{-}\mathbf{C} \qquad (1.2)$$

In polymer chemistry, this reaction is called the *chain growth* (or *chain propagation*) reaction. Reaction 1.2 is very fast; thousands of these reactions occur over a period of a few seconds. This repeated insertion reaction constitutes the process of the growth of a single polymer chain. Each such step can be written in a general form as an increase of the polymer chain length while the chain remains attached to the transition metal atom:

$$[M]^{+}\text{-}(CH_2\text{-}CH_2)_n\text{-}\mathbf{R} + CH_2{=}CH_2 \ -(k_p)\rightarrow \ [M]^{+}\text{-}(CH_2\text{-}CH_2)_{n+1}\text{-}\mathbf{R} \tag{1.3}$$

When an α-olefin $CH_2{=}CH\text{-}\boldsymbol{R}$ is present in a reactor, another chain growth reaction can take place:

$$[M]^{+}\text{-}(CH_2\text{-}CH_2)_n\text{-}\mathbf{R} + CH_2{=}CH\text{-}\boldsymbol{R} \ -(k_p')\rightarrow$$
$$[M]^{+}\text{-}CH_2\text{-}CH\boldsymbol{R}\text{-}(CH_2\text{-}CH_2)_n\text{-}\mathbf{R} \tag{1.4}$$

Reaction 1.3 and Reaction 1.4 are the two principal steps in the formation of ethylene/α-olefin copolymer chains. Ethylene is the monomer of the highest reactivity in catalytic polymerization reactions and Reaction 1.4 always proceeds at a lower rate than Reaction 1.3.

Very infrequently, an ethylene molecule reacts with the growing polymer chain in a different manner than in Reaction 1.3:

$$[M]^{+}\text{-}(CH_2\text{-}CH_2)_n\text{-}\mathbf{R} + CH_2{=}CH_2 \ -(k_t^{E})\rightarrow$$
$$[M]^{+}\text{-}CH_2\text{-}CH_3 + CH_2{=}CH\text{-}(CH_2\text{-}CH_2)_{n-1}\text{-}\mathbf{R} \tag{1.5}$$

Reaction 1.5 is called the *chain transfer reaction to a monomer* (ethylene in this case). This reaction results in the disengagement of the growing polymer chain $[M]^{+}\text{-}(CH_2\text{-}CH_2)_n\text{-}R$ from the active center. The polymer chains $CH_2{=}CH\text{-}(CH_2\text{-}CH_2)_{n-1}\text{-}R$ separated from active centers are called "dead" chains to distinguish them from the growing polymer chains. The new species $[M]^{+}\text{-}CH_2\text{-}CH_3$ formed in Reaction 1.5 still has the $[M]^{+}\text{-}C$ bond and retains the ability to insert ethylene molecules in Reaction 1.2 and grow a new polymer chain.

Reaction 1.5 is the principal chain transfer reaction in ethylene polymerization reactions with chromium oxide and metallocene catalysts. The frequency of Reaction 1.5 strongly depends on temperature: the higher the temperature, the more frequent the reaction is and, as a result, the lower is the average polymerization degree n in "dead" polymer molecules $CH_2{=}CH\text{-}(CH_2\text{-}CH_2)_{n-1}\text{-}\mathbf{R}$.

When titanium-based Ziegler-Natta catalysts are used in ethylene polymerization reactions, Reaction 1.5 occurs very infrequently and the "dead" polymer chains contain, on average, a very large number of ethylene units; the n value can range from 10,000 to 50,000. Polymers with such a high molecular weight are difficult to process; they have found a commercial use only as components of bimodal resins

(Section 1.6 and Section 2.5.2). As mentioned in Section 1.2, a special chemical agent, hydrogen, is nearly always added to ethylene polymerization reactions with Ziegler-Natta catalysts. It hydrogenates the $[Ti]^+$-C bond in the growing polymer chain:

$$[Ti]^+\text{-}(CH_2\text{-}CH_2)_n\text{-}\mathbf{R} + H_2 \text{-}(k_t^H)\rightarrow [Ti]^+\text{-}H + CH_3\text{-}CH_2\text{-}(CH_2\text{-}CH_2)_{n-1}\text{-}\mathbf{R} \quad (1.6)$$

Reaction 1.6 is called the *chain transfer reaction to hydrogen*. The new species with the $[Ti]^+$-H bond remains active and can also insert the double bond of an ethylene molecule, similarly to Reaction 1.2. Reaction 1.6 proceeds with a much higher rate than Reaction 1.5, and the molecular weight of ethylene polymers produced with Ziegler-Natta catalysts in the presence of hydrogen is always significantly lower.

Reaction 1.5 and Reaction 1.6 produce "dead" polymer molecules of different types. One of the end groups in the "dead" molecule formed in Reaction 1.5 is the vinyl double bond, CH_2=CH–. If the last monomer unit in the growing polymer chain is derived from an α-olefin CH_2=CH–***R*** (Reaction 1.4), a different unsaturated end group in a reaction similar to Reaction 1.5 is formed, the vinylidene double bond, CH_2=C***R***–. On the other hand, both types of end groups in "dead" polymer molecules generated in Reaction 1.6 are saturated, CH_3–CH_2– and CH_3–CH***R***–.

Commercial ethylene polymerization reactions are usually carried out at a constant ethylene partial pressure P_E and, therefore, at a constant ethylene concentration C_E. The polymerization rate, the rate of ethylene consumption in Reaction 1.3, is expressed as:

$$R_{pol} = k_p \cdot C^* \cdot C_E \quad (1.7)$$

Here C^* is the concentration of the active centers $[M]^+$–C in the catalyst. The experimental evaluation of the C^* value is very difficult [3, 4]; therefore, the product of two values in Equation 1.7, k_p and C^*, which is called the *effective rate constant*, k_{eff}, is often used as a combined parameter to characterize the catalyst activity.

The average molecular weight MW_{av} of a polyethylene chain formed in Reactions 1.3, 1.5, and 1.6 is equal to the ratio between the probability (or the reaction rate) of the chain growth reaction (Reaction 1.3) and the combined probability of two chain transfer reactions, to a monomer (Reaction 1.5) and to hydrogen (Reaction 1.6):

$$MW_{av} \approx 28 \cdot k_p \cdot C^* \cdot C_E / (k_t^E \cdot C^* \cdot C_E + k_t^H \cdot C^* \cdot C_H) =$$
$$28 \cdot k_p \cdot C_E / (k_t^E \cdot C_E + k_t^H \cdot C_H) \quad (1.8)$$

Here C_H is the hydrogen concentration; k_p, k_t^E, and k_t^H are the rate constants of Reaction 1.3, Reaction 1.5, and Reaction 1.6, respectively; 28 is the molecular weight of an ethylene molecule.

When chromium oxide catalysts or metallocene catalysts are employed in ethylene polymerization reactions, hydrogen is usually not added ($C_H = 0$). In this case, Equation 1.8 is simplified to $MW_{av} = 28 \cdot k_p/k_t^E$; that is, the average molecular weight of polyethylene produced in the absence of hydrogen does not depend on the ethylene concentration C_E in a reactor. On the other hand, most ethylene polymerization reactions with Ziegler-Natta catalysts are carried out at a significant hydrogen pressure when the $k_t^E \cdot C_E$ term in Equation 1.8 is much lower than the $k_t^H \cdot C_H$ term. Consequently, $MW_{av} \approx 28(k_p/k_t^H) \cdot (C_E/C_H)$; that is, the average molecular weight of polyethylene resins produced with these catalysts depends on the ratio between concentrations of ethylene and hydrogen.

All major types of olefin polymerization catalysts, including supported Ziegler-Natta catalysts, titanium-based catalysts used in solution polymerization processes, and chromium oxide catalysts, contain several types of active centers. The centers differ in three respects:

1. Different active centers have different kinetic parameters, different values of the rate constants k_p, k_t^E, and k_t^H. As a result, different active centers produce polymer molecules of a different molecular weight (Equation 1.8). Ethylene polymers produced with all these catalysts always have a broad molecular weight distribution. This effect is described in Section 1.5 and Section 1.6.
2. Different active centers have a different ability to copolymerize ethylene with α-olefins. This feature of the active centers is described in Section 1.8. Although all the commercially manufactured ethylene/α-olefin copolymers are produced under strictly constant [ethylene]:[α-olefin] ratios, different copolymer molecules in the reaction products have a different composition. Some active centers exhibit a very poor copolymerization ability. These centers produce copolymer molecules with a low α-olefin content. Other active centers copolymerize α-olefins with ethylene quite well and produce copolymer molecules with a high α-olefin content. In contrast, LLDPE resins produced with metallocene and some post-metallocene catalysts are usually compositionally uniform; all macromolecules in these products have approximately the same α-olefin content.
3. Different active centers have different kinetic stability. This difference is especially pronounced in ethylene/α-olefin copolymerization reactions with Ziegler-Natta catalysts. The centers that do not copolymerize α-olefins with ethylene well are relatively stable. If the copolymerization reaction is carried at 70–80 °C, the reactivity of these centers does not change for a period of 2 to 3 hours. The centers that copolymerize α-olefins with ethylene well are usually less stable; at 70–80 °C they decompose after 30 to 40 minutes.

Two more chemical reactions are essential for understanding structural, rheological, and mechanical properties of some polyethylene resins. The first such reaction is the formation of long-chain branches in polyethylene chains. The "dead" polymer chain formed in Reaction 1.5, CH_2=CH-(CH_2-CH_2)$_{n-1}$-**R**, has the same vinyl double bond CH_2=CH- as any α-olefin molecule CH_2=CH-***R*** employed in ethylene/α-olefin copolymerization reactions. The difference between these two α-olefin molecules is merely the size of the alkyl group ***R*** attached to the vinyl double bond; this alkyl group is small in α-olefin molecules used in the copolymerization reactions (from C_2H_5 to C_6H_{13}) but it can be very large in "dead" polymer molecules. It has been proven experimentally that when ethylene polymerization reactions are catalyzed by some metallocene and chromium oxide catalysts, such "dead" polymer molecules containing the vinyl double bond participate in copolymerization reactions with ethylene (Reaction 1.4) [32–34]. Copolymerization of ethylene with the "dead" polymer molecules produces long side-groups (long-chain branches) in polyethylene molecules. Although the long-chain branches are always present in a very low concentration, they greatly affect the rheology of the polymer melt (Section 2.5.3).

The second peculiar chemical reaction accompanying the formation of polymer chains was discovered in ethylene polymerization reactions with post-metallocene catalysts based on late-period transition metals, in particular, α-diimine complexes of Ni and Pd [3, 4, 35, 36]. Active centers in these catalysts have a tendency to form isomerized polymer chains. This reaction is called "a chain walking reaction". It accompanies the chain growth reaction (Reaction 1.2) and can be schematically represented as a shift of the metal atom position along a crowing polyethylene chain from [M]+-CH_2-CH_2-CH_2-CH_2-CH_2-R to [M]+-CH(CH_3)-CH_2-CH_2-CH_2-R. The chain growth steps in Reaction 1.2 continue after the shift and, as a result, methyl branches in polyethylene chains are formed, R′-CH_2-CH_2-CH(CH_3)-CH_2-CH_2-CH_2-R. The content of the methyl branches ranges from ~4 to over 7 mol % depending on the reaction temperature [35, 36]. Other types of branches, from ethyl to hexyl, are also formed in the chain-walking reactions but with lower amounts, from 0.1 to ~1 mol %. The formation of the branches leads to a significant depression of the resin's melting point – see Chapter 3.

1.5 Molecular Weight Distribution of Polyethylene and Methods of Its Analysis

A typical active center $[M]^+$-C in a polymerization catalyst has the life-span from several minutes (if the polymerization reaction is carried out at a very high temperature) to several hours. During this period, every center produces several thousand polymer molecules. The formation of a single macromolecule containing from 2,000 to 10,000 monomer units typically takes from several seconds to a minute. The length of time each macromolecule grows before disengaging from the active center in Reaction 1.5 or Reaction 1.6 is determined by pure statistics; the majority of the "dead" macromolecules are relatively short, but others are very long.

Flory distribution function: The simplest case of the molecular weight distribution describes relative fractions of macromolecules of different length, that is, macromolecules containing different number of ethylene units, n. It is usually referred to as the Flory-Schulz distribution function [3, 37–40]. This function applies to polymerization processes with the following characteristics:

1. Catalytic polymerization reactions in which all the active centers have the same kinetic characteristics and the same reactivity.
2. Polymerization reactions in which the probability of the polymer chain separation from an active center in Reaction 1.5 or Reaction 1.6 does not depend on the number n of monomer units attached to the transition metal M.
3. Polymerization reactions in which concentrations of all the reactants remain constant. The last condition is typical for commercial polymerization processes, which are usually carried out in the continuous mode, when the concentrations of the reactants are kept constant and are closely monitored.

The theory of the molecular weight distribution states that if a polymerization reaction produces macromolecules of a high molecular weight, the distribution function of polymer chains, $F_{\text{number}}(n)$, is given by the following equation [3, 37–43]:

$$F_{\text{number}}(n) = \mathbf{n}_{\text{av}}^{-1} \cdot \text{Exp}\left(-n/\mathbf{n}_{\text{av}}\right) \tag{1.9}$$

The $F_{\text{number}}(n)$ value in Equation 1.9 is the fraction of polymer chains containing n monomer units and $\mathbf{n}_{\text{av}}$ is the average polymerization degree of the produced mixture of polymer chains. In the case of polyethylene resins, a polymer chain containing n monomer units has the molecular weight of $28 \cdot n$, and the number-average molecular weight of the polymer described by the Flory-Schulz equation is $M_{\text{n}} = 28 \cdot \mathbf{n}_{\text{av}}$. The M_{n} value has the same meaning as the kinetically defined MW_{av} value in Equation 1.8. The Flory-Schulz distribution signifies that a polymerization reaction is characterized by a single $\mathbf{n}_{\text{av}}$ value and that the $\mathbf{n}_{\text{av}}$ value is constant in the course of a given polymerization reaction.

The $F_{\text{number}}(n)$ function states that the shorter a given polymer chain (the lower the n value) the higher is the fraction of such chains in the polymer mixture. However, for a polymer chemist or an engineer interested in polymer properties, the distribution function that matters is not the fraction of particular polymer chains in a given polymer product but the *weight fraction* of polymer material in these chains. The weight distribution function, $F_{\text{weight}}(n)$, represents the normalized weight fraction of polymer chains consisting of n monomer units:

$$F_{\text{weight}}(n) = n \cdot F_{\text{number}}(n) \Big/ \int \left[n \cdot F_{\text{number}}(n)\right] \cdot dn = \left(n/\mathbf{n}_{\text{av}}^{2}\right) \cdot \text{Exp}\left(-n/\mathbf{n}_{\text{av}}\right) \tag{1.10}$$

The weight-average polymerization degree $\mathbf{n}_{weight}^{av}$ of such a polymer is equal to $2 \cdot \mathbf{n}_{av}$; the weight-average molecular weight is $M_w = 56 \cdot \mathbf{n}_{av}$. Polymer chemistry widely uses the ratio of M_w and M_n values to characterize the width of the molecular weight distribution of a given polymer. In the case of polymers with a molecular weight distribution described by the Flory-Schulz function (Equation 1.9), the width of the molecular weight distribution, $M_w/M_n = \mathbf{n}_{weight}^{av}/\mathbf{n}_{av}$, is equal to 2.

Figure 1.1 shows an example of the shapes of two functions, $F_{number}(n)$ and $F_{weight}(n)$, for an ethylene polymer with $\mathbf{n}_{av} = 1{,}000$ ($M_w = 56{,}000$). The $F_{number}(n)$ function is an exponentially decreasing curve whereas the $F_{weight}(n)$ function has a maximum at $n = \mathbf{n}_{av}$.

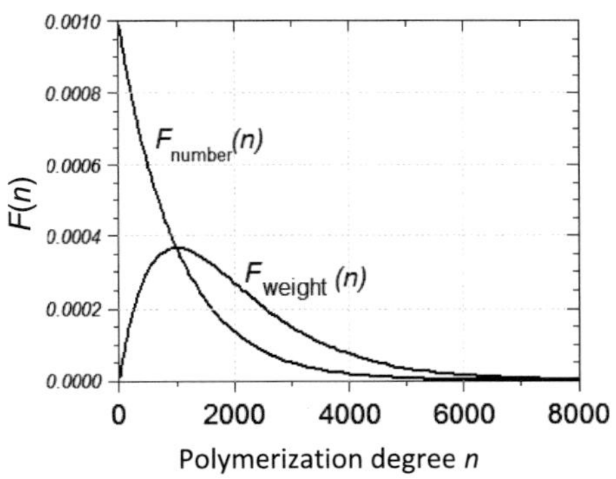

Figure 1.1
Functions $F_{number}(n)$ and $F_{weight}(n)$ for polymer with $\mathbf{n}_{av} = 1{,}000$

Experimental measurement methods of molecular weight distribution: At the present time, the gel permeation chromatographic (GPC) method is the universal technique for the measurement of average molecular weights and the molecular weight distribution of polyethylene resins. Another name of this technique is the size-exclusion chromatography or SEC [3, 41]. Several instruments for the GPC analysis are commercially produced. To carry out a GPC analysis, a dilute solution of a resin in a good solvent (usually *ortho*-dichlorobenzene or 1,2,4-trichlorobenzene) passes through a set of columns filled with a cross-linked polystyrene gel. To assure complete dissolution of high molecular weight resins in the solvents, the analysis is carried out at a high temperature, from 110 to 120 °C. Due to differences in diffusion rates of polymer molecules of a different size through layers of the swelled gel, the average residence time of a particular polymer molecule in the columns is in the reciprocal dependence to its molecular weight: the largest molecules leave the columns rapidly whereas shorter molecules meander through the gel for a significant period of time. The concentration of polymer molecules exiting the last column is measured with a highly sensitive detector.

The most advanced variants of GPC instruments currently on the market employ two detectors and carry out the second analysis of the polymer solution leaving the

GPC columns in addition to measuring the polymer content in the solution. The first of the techniques was developed for the analysis of ethylene/α-olefin copolymers [44–47]. It measures the content of α-olefin in the eluted copolymer molecules by the Fourier-transform Infrared (FTIR) method. Another double-detector method is used to analyze the molecular weight distribution of resins containing long-chain branches in LDPE and in some metallocene resins [48]. In this case, the second detector is an online viscometer.

Flory distribution function in GPC coordinates: In order to describe the shape of the GPC curve of a polymer with the molecular weight distribution given by Equation 1.10, one has to take into account a peculiar nature of the GPC analysis: a nearly perfect linear correlation exists between the time a given polymer molecule spends in a GPC column and the logarithm of its molecular weight (or its polymerization degree n). Employing this relationship, the expression for the Flory weight distribution function, $F_{weight}(n)$ in Equation 1.10, is transformed into the following expression in the GPC coordinates [3, 42, 43]:

$$F_{weight}\left[\log(n)\right] = \mathbf{n}_{av}^{-2} \cdot \mathrm{Exp}\left\{2 \cdot \ln(10) \cdot \log(n) - \mathbf{n}_{av}^{-1} \cdot \mathrm{Exp}\left[\ln(10) \cdot \log(n)\right]\right\} \quad (1.11)$$

where $\ln(10) \approx 2.3$. The $F_{weight}[\log(n)]$ function in Equation 1.11 represents the same weight distribution function of polymer chains as the $F_{weight}(n)$ function in Equation 1.10 but with a different variable, $\log(n)$, instead of n. As a rule, GPC curves of polyethylene resins are plotted using a slightly different abscissa, $\log(MW) = \log(28 \cdot n)$ rather than $\log(n)$.

Figure 1.2 shows the GPC curve of a Flory-distributed ethylene polymer with $M_w = 56{,}000$. The curve is calculated with Equation 1.11 in the coordinates "$F_{weight}[\log(n)]$ as a function of $\log(MW) = \log(28 \cdot n)$"; it has a sharp maximum and it is asymmetric: its high molecular-weight side has a steeper slope than the low molecular-weight side. Although the GPC curve in Figure 1.2 and the distribution curve $F_{weight}(n)$ in Figure 1.1 have different shapes, they represent the molecular weight distribution of the same polymer with $M_w = 56{,}000$ and $\mathbf{n}_{av} = 1{,}000$ but in different coordinates.

The $F_{weight}[\log(n)]$ function in Equation 1.11, as well as the $F_{weight}(n)$ function in Equation 1.10, contains only one parameter, the number-average polymerization degree $\mathbf{n}_{av}$. The shape of the curve in Figure 1.2 (the shape of the Flory function in the GPC coordinates) does not depend on any adjustable parameters for the width of the peak. The maximum of the Flory function in the GPC coordinates is positioned at $\log(n)_{max} = \log(2 \cdot \mathbf{n}_{av})$, that is, at $n_{max} = n_{weight} = 2 \cdot \mathbf{n}_{av}$ or at $MW^{max} = M_w$.

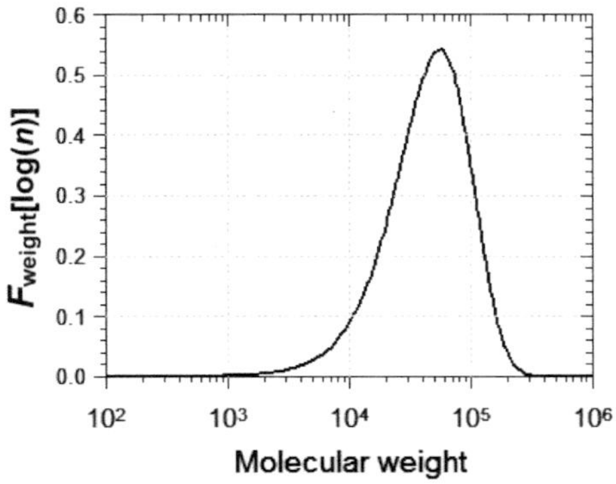

Figure 1.2
GPC curve of polyethylene resin with single-Flory molecular weight distribution calculated with Equation 1.11

The Flory function can be readily applied to describe the molecular weight distribution of any resin produced in a catalytic polymerization reaction under stationary conditions. In the general case, a polyethylene resin is characterized by the following three sets of parameters:

1. The number of Flory components, **j**. The **j** value can vary from one in the case of some metallocene resins to as high as 7 to 8 in the case of resins produced with chromium oxide catalysts and resins with a bimodal molecular weight distribution.
2. Each Flory component has the weight-average molecular weight $M_{w,j}$, the number-average molecular weight $M_{n,j} = M_{w,j}/2$, and the respective average polymerization degree $\mathbf{n}_{av,j} = M_{w,j}/56$.
3. The content of each Flory component in the mixture is given by its weight fraction FR_j.

The combined number distribution function for such a polymer mixture with respect to the polymerization degree n is (similar to Equation 1.9):

$$F_{number}(n)^{total} = \sum FR_j \cdot \mathbf{n}_{av,j}^{-1} \cdot \mathrm{Exp}\left(-n/\mathbf{n}_{av,j}\right) \quad (1.12)$$

The combined weight distribution function (similar to Equation 1.10) is:

$$F_{weight}(n)^{total} = n \cdot \sum FR_j \cdot \mathbf{n}_{av,j}^{-2} \cdot \mathrm{Exp}\left(-n/\mathbf{n}_{av,j}\right) \quad (1.13)$$

The number-average molecular weight of the multi-component mixture, M_n^{av}, is:

$$M_n^{av} = \left[\sum FR_j / M_{n,j}\right]^{-1} \quad (1.14)$$

The weight-average molecular weight of the multi-component mixture, M_w^{av}, is:

$$M_w^{av} = \left[\sum FR_j \cdot 2 \cdot M_{n,j}\right]^{-1} \quad (1.15)$$

The width of the molecular weight distribution of the mixture, $(M_w/M_n)^{av}$, is:

$$\left(M_w/M_n\right)^{av} = M_w^{\ av}/M_n^{\ av} \tag{1.16}$$

Several examples in the next section demonstrate that a GPC curve of any polyethylene resin, however broad and complex, can be represented in a satisfactory manner as a combination of several Flory curves in the GPC coordinates, each described by Equation 1.11 and characterized by a single parameter, M_w.

1.6 Examples of Molecular Weight Distribution of Polyethylene Resins

1.6.1 Polyethylene Resins Produced with Metallocene Catalysts

Polyethylene resins with the narrowest molecular weight distribution, M_w/M_n ~ 2.0, can be produced only with single-site soluble metallocene catalysts and only at a very high [MAO]:[metallocene complex] ratio, at $[Al]:[M]_{molar} > 2{,}000$ to 3,000. One such example is given in Figure 1.3. It shows the GPC curve of an LLDPE resin produced with a soluble metallocene catalyst at the [MAO]:[metallocene complex] ratio of ~ 10,000 (dots) [42]. The solid line represents the GPC curve of a single Flory component; it was calculated with Equation 1.11.

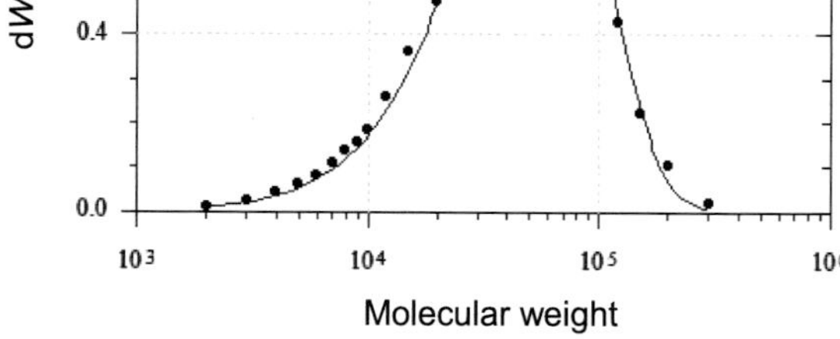

Figure 1.3 Experimental GPC curve of ethylene/1-hexene copolymer with C_M^{copol} = 2.0 mol % produced with metallocene catalyst (points). The line is a single-Flory GPC curve calculated with Equation 1.11

Resins with such a narrow molecular weight distribution represent a rare exception. Even the metallocene resins prepared with soluble catalysts at low [MAO]:[metallocene complex] ratios, and all metallocene resins prepared with supported catalysts always have a noticeably broader molecular weight distribution. As an example, Figure 1.4 shows the GPC curve of a VLDPE resin (ethylene/1-hexene copolymer with C_M^{copol} = 17 mol %) produced with a soluble metallocene catalyst at the [MAO]:[metallocene complex] ratio of ~1,000. This material consists of two Flory components with average molecular weights M_w of 1,700 and 4,100, respectively, in a ~0.7:1 weight ratio [42].

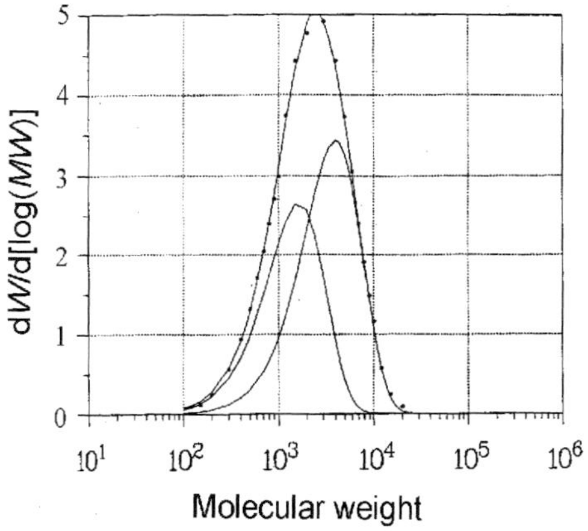

Figure 1.4
Experimental GPC curve of VLDPE resin (ethylene/1-hexene copolymer with C_M^{copol} ~17 mol %) produced with metallocene catalyst (points). The lines are calculated GPC curves of individual Flory components and their sum

1.6.2 Polyethylene Resins Produced with Supported Ziegler-Natta Catalysts

Figure 1.5 shows the GPC curve of an HDPE resin produced with a supported Ziegler-Natta catalyst at 80 °C and its resolution into Flory components. Four components with M_w values ranging from ~16,000 to ~340,000 are present in the product. The weight-average molecular weight of the resin calculated with Equation 1.15 is 133,000 and the width of its molecular weight distribution, $(M_w/M_n)^{av}$, calculated with Equation 1.16 is 6.0. A very small low molecular weight component (~1.5%) with M_w ~3,000 is possibly an artifact of GPC recording conditions or the computer peak resolution program.

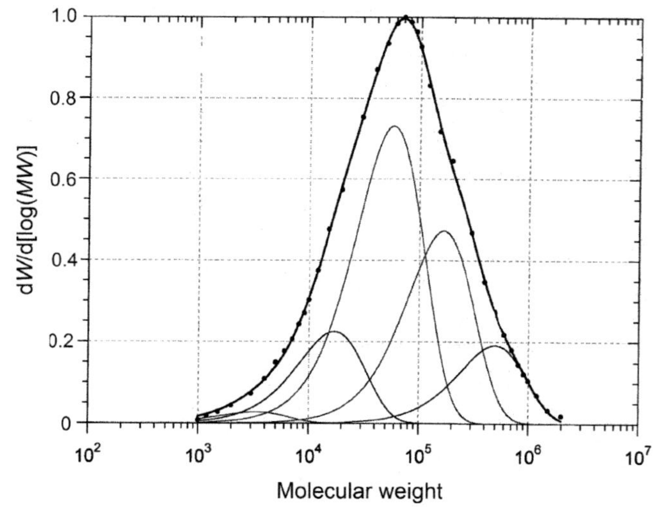

Figure 1.5 Experimental GPC curve of HDPE resin produced with supported Ziegler-Natta catalyst at 80 °C (points) and its resolution into Flory components

Table 1.2 gives parameters M_w and FR_j of each Flory component in this resin. The numbering of Flory components used in the table, from I to IV in the order of increasing molecular weight, as well as numbering of Flory components in all the following tables, is arbitrary.

Table 1.2 Flory Components in HDPE Resin Produced with Supported Ziegler-Natta Catalyst

Flory component (j)	Molecular weight ($M_{w,j}$)	Content (FR_j), %
I	15,700	14
II	47,900	45
III	124,400	29
IV	342,900	12
Total resin	133,000	100

Figure 1.6 shows the GPC curve of an LLDPE resin (ethylene/1-hexene copolymer with C_M^{copol} = 3.5 mol) produced with a supported Ziegler-Natta catalyst at 85 °C. The product is also a mixture of four Flory components [49]. The distribution of copolymer molecules with respect to their molecular weight in these LLDPE resins is also quite broad, similarly to HDPE resins produced with catalysts of the same type (Figure 1.5) [42, 50]. The width of the molecular weight distribution $(M_w/M_n)^{av}$ of the copolymer is 4.0.

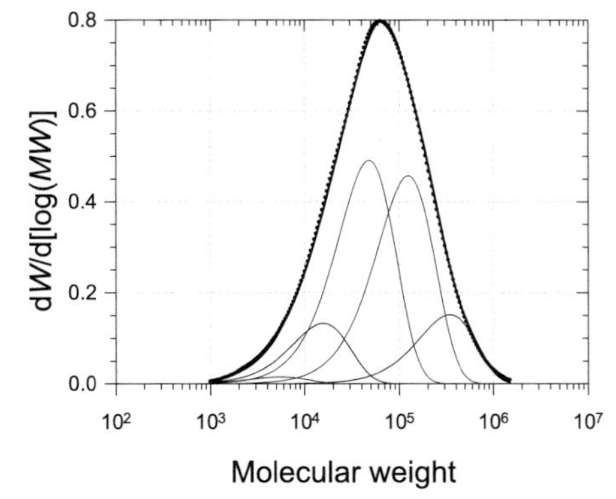

Figure 1.6
Experimental GPC curve of LLDPE resin (ethylene/1-hexene copolymer with C_M^{copol} = 3.5 mol %) produced with supported Ziegler-Natta catalyst and its resolution into Flory components

Table 1.3 gives parameters of each Flory component in the copolymer mixture. Detailed kinetic analysis of ethylene/1-hexene copolymerization reactions with supported Ziegler-Natta catalysts showed [50–52] that Flory components in the copolymers differ not only in the average molecular weight but also in the 1-hexene content. As Table 1.3 shows, the components with the lowest molecular weight (components I and II) have the highest 1-hexene content whereas the components with the highest molecular weight (components III and IV) have the lowest 1-hexene content. The existence of the differences in the copolymer composition was confirmed with several crystallization fractionation studies described in Section 1.8 (see Table 1.7).

Table 1.3 Flory Components in LLDPE Resin Produced with Supported Ziegler-Natta Catalyst

Flory component (j)	Molecular weight ($M_{w,j}$)	Content (FR_j), %	C_M^{copol}, mol %
I	15,700	11	8 to 10
II	47,900	39	~4
III	124,400	37	0.6 to 0.8
IV	342,900	12	0.3 to 0.4
Total resin	109,000	100	3.5

1.6.3 Polyethylene Resins Produced with Chromium Oxide Catalysts

Figure 1.7 shows the GPC curve of a typical HDPE resin prepared with a chromium oxide catalyst. The molecular weight distribution of such resins is always very broad; seven Flory components are needed to represent this GPC curve. Table 1.4 lists molecular weights of Flory components in an HDPE resin produced at 90 °C.

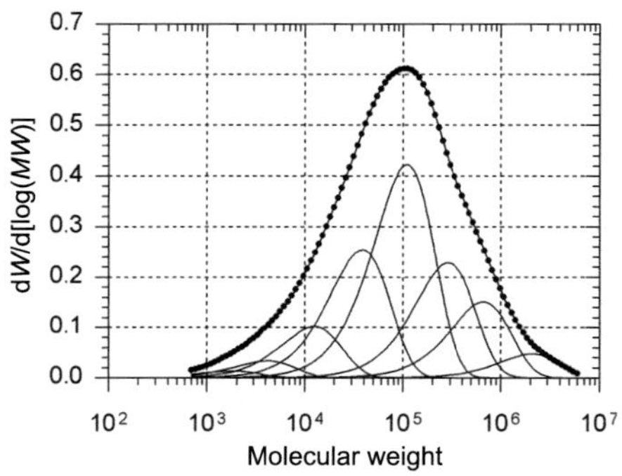

Figure 1.7
Experimental GPC curve of HDPE resin produced with chromium oxide catalyst (points) and its resolution into Flory components

The range of molecular weights of the Flory components in the polymer mixture is very wide, from ~3,000 to nearly 2,000,000. As a result, the width of the molecular weight distribution, the $(M_w/M_n)^{av}$ ratio, is also high, ~12. Obviously, the *average* molecular weight of a resin with such a broad molecular weight distribution is a poor predictor of its physical and mechanical properties.

Table 1.4 Flory Components in HDPE Resin Produced with Chromium Oxide Catalyst

Flory component (j)	Molecular weight ($M_{w,j}$)	Content (FR_j), %
I	3,200	3
II	12,100	10
III	38,500	25
IV	96,700	25
V	228,000	21
VI	598,900	14
VII	1,848,000	3
Total resin	227,000	100

1.6.4 Polyethylene Resins with Bimodal Molecular Weight Distribution

Polyethylene resins with a bimodal molecular weight distribution are mixtures of two fractions, one with a very high molecular weight and another with a very low molecular weight. The characteristic feature of such resins is the shape of their GPC curves, which consist of two clearly identifiable components. Two types of bimodal resins are produced on the industrial scale. Resins of the first type are produced with bimetallic catalysts. One of the catalyst components in a bimetallic catalyst is a metallocene or a post-metallocene catalyst, which produces the low molecular weight fraction of the bicomponent resin, and the second catalyst component is a Ziegler-Natta catalyst, which produces the high molecular weight fraction. The second type of bimodal resins is produced with a single Ziegler-Natta catalyst in two reactors connected in a series and operating under different reaction conditions (Section 1.2).

Figure 1.8 shows the GPC curve of an HDPE resin with a bimodal molecular weight distribution prepared with a bimetallic metallocene/Ziegler-Natta catalyst and the GPC curves of all its Flory components. The two Flory components with the lowest molecular weights are produced by the metallocene catalyst; together they account for ~35% of the total resin weight. The four high molecular weight Flory components are produced by the titanium-based Ziegler-Natta catalyst.

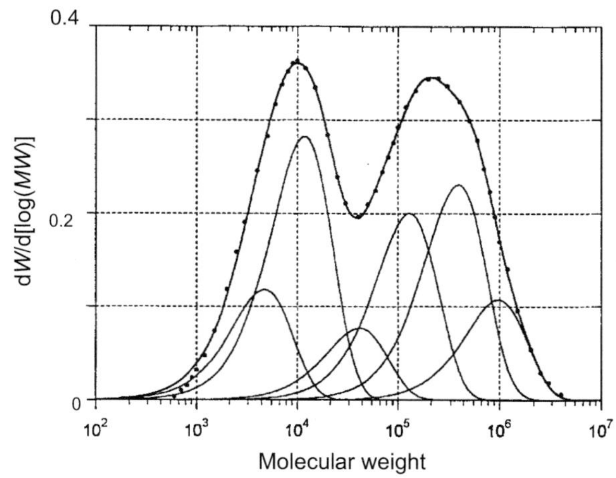

Figure 1.8 Experimental GPC curve of bimodal HDPE resin produced with bimetallic metallocene/Ziegler-Natta catalyst (points) and its resolution into Flory components

Table 1.5 gives an example of molecular weight characteristics of both fractions in a pipe-grade HDPE resin prepared with a bimetallic catalyst. The resin fraction produced with the metallocene component consists of two Flory components with M_w ~4,000 and ~11,000. Its molecular weight distribution is narrow, $(M_w/M_n)^{av} = 3$; and its weight-average molecular weight is very low, ~10,000. Mechanical properties this waxy material, if synthesized separately, would be very poor. The principal function of this fraction in the bimodal polymer mixture is to serve as a low-viscosity diluent for the high molecular weight material. The resin fraction produced by the Ziegler-Natta component of the bimetallic catalyst consists of four Flory components, the same as in the HDPE resin shown in Figure 1.5. Molecular weights of all the Flory components in this fraction are high; they range from ~40,000 to 1,000,000. This material provides the necessary strength for articles manufactured from bimodal resins, but, if produced separately, it would form a very viscous, unprocessable melt. As in the case of the HDPE resin produced with a chromium oxide catalyst (Figure 1.7), the value of the average molecular weight in such complex mixtures does not reflect end-use properties of the resin.

Table 1.5 Components of Film-Grade HDPE Resin Produced with Bimetallic Metallocene/Ziegler-Natta Catalyst

Resin	Number of Flory components	M_w^{av}	$(M_w/M_n)^{av}$	Content, %
Total polymer	6	187,000	20	100
Fraction produced with Ziegler-Natta catalyst	4	275,800	4.7	67
Fraction produced with metallocene catalyst	2	10,200	3.0	33

Figure 1.9 shows the GPC curve of another type of an HDPE resin with a bimodal molecular weight distribution. This resin was prepared with the use of a single Ziegler-Natta catalyst in two slurry reactors connected in a series. The concentration of the chain transfer agent (hydrogen) was very different in the two reactors, and as a result, the final resin is a mixture of two fractions; one with a very high molecular weight and another with a very low molecular weight. Parameters of the total resin and its two components are listed in Table 1.6. The GPC curve of this resin is the overlap of two nearly identical curves. Each curve consists of four Flory components, the same as in Figure 1.5 and Figure 1.8. Due to differences in the synthesis conditions, the components are shifted by a factor of ~19 in terms of their average molecular weights. Although each resin fraction has a medium-broad molecular weight distribution, $(M_w/M_n)^{av}$ ~4, the combined molecular weight distribution of the bimodal resin is very broad, $(M_w/M_n)^{av}$ >25.

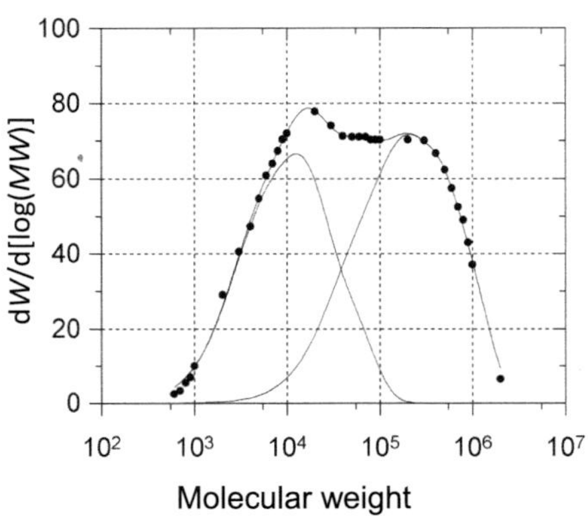

Figure 1.9
Experimental GPC curve of bimodal HDPE resin produced with supported Ziegler-Natta catalyst in two slurry reactors connected in series (points) and its resolution into curves of two fractions.

Table 1.6 Components of Film-Grade HDPE Resin Produced with Ziegler-Natta Catalyst in Two-Reactor Process

Resin	M_w^{av}	$(M_w/M_n)^{av}$	Content, %
Total polymer	248,400	26.4	100
Low molecular weight fraction	17,800	3.9	45
High molecular weight fraction	343,500	3.9	55

■ 1.7 Copolymer Statistics and Its Application to Description of LLDPE and VLDPE Resins

All commercially produced LLDPE and VLDPE resins are copolymers of ethylene with α-olefins, 1-butene, 1-hexene, 1-octene, or 4-methyl-1-pentene. The copolymers differ by the amount of α-olefin (Table 1.1), but this amount is always relatively low. For example, all LLDPE resins, whatever their manufacturer, the catalyst employed, or the reactor process is, contain between 2.5 and 3.5 mol % of α-olefin.

Any copolymer macromolecule containing ethylene monomer units, $-CH_2-CH_2-$, and monomer units of an α-olefin, $-CH_2-CH\boldsymbol{R}-$, can be viewed as consisting of sets of monomer blocks (monomer sequences):

Blocks of n ethylene units, $-CH_2-CH\boldsymbol{R}-(CH_2-CH_2)_n-CH_2-CH\boldsymbol{R}-$ or $M-(E)_n-M$,

Blocks of m α-olefin units, $-CH_2-CH_2-(CH_2-CH\boldsymbol{R})_m-CH_2-CH_2-$ or $E-(M)_m-E$.

Each block is flanked by two monomer units of the opposite kind: the ethylene blocks are $M-(E)_n-M$ and the α-olefin blocks are $E-(M)_m-E$. The number of monomer units in the blocks, n or m, can vary starting from one (isolated monomer units, sequences M–E–M or E–M–E) to any large number.

Because the content of α-olefin in LLDPE resins is low, the blocks of α-olefin units are all short; most m values are either 1 or, rarely, 2 or 3. The crystallinity level of all LLDPE resins and the value of their melting point are entirely determined by the presence of relatively long ethylene sequences. The ethylene/α-olefin copolymers are random, that is, the ethylene and the α-olefin units are positioned along copolymer chains in a strictly statistical manner. Statistical expressions for calculating the contents of ethylene units in blocks $M-(E)_n-M$ with different n values are well known [2, 3]. The principal starting parameter in the calculations is the molar percent of α-olefin in a copolymer C_M^{copol}. The higher the C_M^{copol} value the shorter (on average) are the ethylene blocks and the lower is the fraction of ethylene units in the long blocks.

Two statistical functions are usually used to describe ethylene blocks in random copolymers [2, 3, 53]:

1. The fraction of ethylene units in blocks containing n units (that is, in blocks M-$(E)_n$-M) normalized to the total molar content of ethylene in the copolymer, $Fraction_E^{copol} = (1 - C_M^{copol}/100)$:

$$\delta(E)_n = n \cdot \left(C_M^{copol}/100\right)^2 \cdot [1 - \left(C_M^{copol}/100\right)]^{n-1} \tag{1.17}$$

2. The fraction of ethylene units in the sum of all long ethylene blocks starting with the block containing n units; that is, the fraction of ethylene units in the sum of blocks M-$(E)_n$-M, M-$(E)_{n+1}$-M, M-$(E)_{n+2}$-M, and so on, also normalized to the total molar content of ethylene in the copolymer. This fraction is given by the expression:

$$\sum(E)_n = \left[\left(C_M^{copol}/100\right) \cdot (n-1) + 1\right] \cdot \left\{\left[1 - \left(C_M^{copol}/100\right)\right]\right\}^{n-1} \tag{1.18}$$

Two examples of the $\delta(E)_n$ function are shown in Figure 1.10. One example is for the copolymer with C_M^{copol} = 3 mol %, which is a typical composition of an LLDPE resin. Another example is for a copolymer with C_M^{copol} = 8 mol %, a typical composition of an ethylene plastomer. It is obvious that most ethylene units in both resins are positioned in long sequences, a fact that accounts for their polyethylene-type crystallinity. The distribution of ethylene units strongly depends on the copolymer composition: the lower the C_M^{copol} value, the higher is the fraction of ethylene units in long (crystallizable) blocks. In the case of the copolymer with C_M^{copol} = 3 mol %, the highest fraction of ethylene units is in blocks ranging from M-$(E)_{30}$-M to M-$(E)_{35}$-M, whereas the most abundant blocks in the copolymer with C_M^{copol} = 8 mol % are shorter, from M-$(E)_{12}$-M to M-$(E)_{14}$-M.

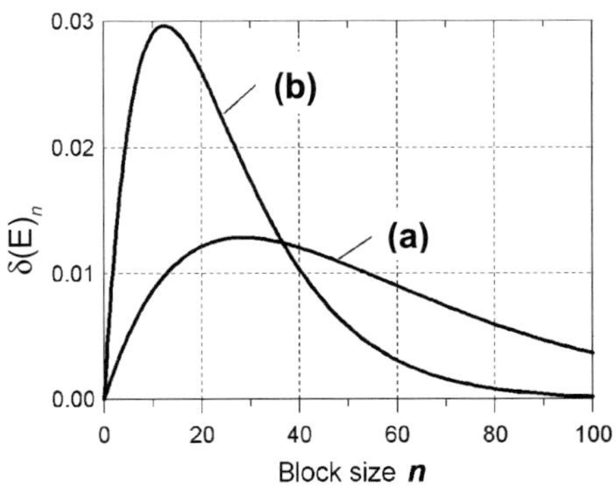

Figure 1.10
Distribution of ethylene units in blocks of different size in ethylene/α-olefin copolymers. (a) LLDPE resin with C_M^{copol} = 3 mol %, (b) VLDPE resin with C_M^{copol} = 8 mol %

The degree of crystallinity of ethylene/α-olefin copolymers is usually measured by the X-ray method or by the differential scanning calorimetry method (Chapter 4). The degree of crystallinity can be approximately estimated by assuming that the resin crystallinity is the function of the fraction of ethylene units in the sum of sufficiently long ethylene blocks $\Sigma(E)_n$ in Equation 1.18; see Section 4.1.

1.8 Compositional Uniformity of Polyethylene Resins

The term *compositional uniformity* is used to describe differences in the composition of different copolymer molecules in polyethylene resins of LLDPE and VLDPE grades. Some of these resins are compositionally uniform. Any copolymer molecule in such resins, taken at random, contains about the same fraction of α-olefin units, C_M^{copol} (the same number of branches), as any other copolymer molecule. The majority of LLDPE and VLDPE resins produced with metallocene catalysts belong to this category. In contrast, all LLDPE and VLDPE resins produced with Ziegler-Natta catalysts and with chromium oxide catalysts are mixtures of copolymer molecules with very different contents of α-olefins. Some macromolecules in these mixtures contain a very small fraction of α-olefin units, while other copolymer molecules contain a relatively large number of the units.

At the present time, two complementary techniques for the analysis of compositional uniformity exist: analytical temperature-rising elution fractionation (TREF) [54–59] and crystallization fractionation (CRYSTAF) [60–67]. Both techniques exploit the same principle: copolymer molecules of different composition crystallize from solution at different temperatures. The first step in both methods is dissolution of a small resin sample in a suitable solvent at a high temperature. The two most often used solvents are *ortho*-dichlorobenzene and 1,2,4-trichlorobenzene. The dissolution temperature is 130 to 140 °C. The well-mixed hot solution is slowly cooled, at a rate of several °C/hour, resulting in slow crystallization of the polymer. In the CRYSTAF method, the concentration of the polymer remaining in solution as a function of temperature is monitored with an infrared detector. In the TREF method, the polymer is completely crystallized on an inert inorganic carrier and then it is slowly re-dissolved in a fresh batch of solvent at a gradually increasing temperature. The concentration of the dissolved polymer is monitored as a function of temperature. Fully automated TREF and CRYSTAF instruments are commercially available; they afford a very thorough resolution of copolymer mixtures into compositionally uniform fractions [60, 61]. Special curve-resolution techniques were developed to identify individual compositionally uniform components in complex TREF and CRYSTAF curves [43, 66].

Several examples of the resolving ability of TREF and CRYSTAF techniques and the results of computer simulation of complex multi-peak fractionation curves are shown in Figure 1.11 to Figure 1.14. Figure 1.11 shows the CRYSTAF curve of an artificial 1:1 mixture of two compositionally uniform LLDPE resins with a different comonomer content (ethylene/1-butene copolymers with C_M^{copol} = 6 and 17 mol %) and its resolution into the peaks of individual components [66].

Several examples of the resolving ability of TREF and CRYSTAF techniques and the results of computer simulation of complex multi-peak fractionation curves are shown in Figure 1.11 to Figure 1.14. Figure 1.11 shows the CRYSTAF curve of an artificial 1:1 mixture of two compositionally uniform LLDPE resins with a different comonomer content (ethylene/1-butene copolymers with C_M^{copol} = 6 and 17 mol %) and its resolution into the peaks of individual components [66].

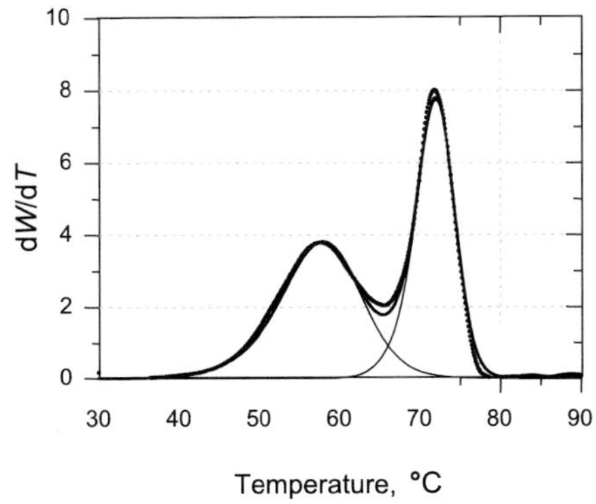

Figure 1.11
CRYSTAF curve of a mixture of two compositionally uniform LLDPE resins with different comonomer content and its resolution into peaks of individual components

Figure 1.12 shows the TREF curve of an LLDPE resin, an ethylene/1-butene copolymer with C_M^{copol} = 2.8 mol % produced with a supported Ziegler-Natta catalyst at 80 °C, and the curve resolution into compositionally uniform fractions [57]. Table 1.7 lists the maximums of the TREF elution peaks (T_{max}) and parameters of the crystallized/eluted fractions. The fractions are marked A to F in the order of decreasing crystallization temperature.

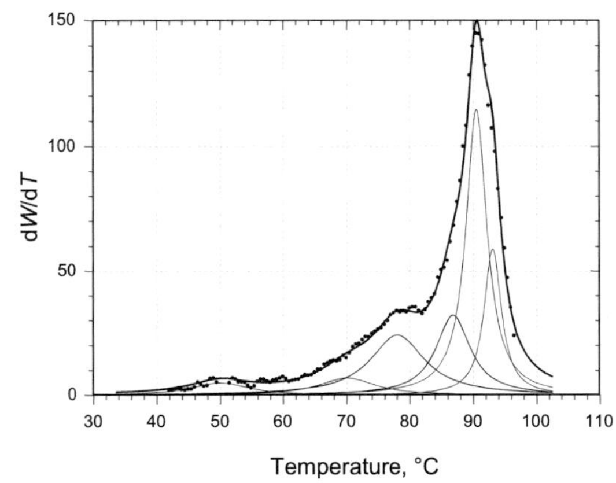

Figure 1.12
TREF curve of LLDPE resin produced with supported Ziegler-Natta catalyst (ethylene/1-butene copolymer, C_M^{copol} = 2.8 mol %) and its resolution into compositionally uniform components

Table 1.7 Compositionally Uniform Components in LLDPE Resin Produced with Supported Ziegler-Natta Catalyst (Ethylene/1-Butene Copolymer with C_M^{copol} = 2.8 mol %); TREF Analysis

Fraction:	A	B	C	D	E	F
T_{max}, °C:	93.1	90.5	86.8	78.0	70.0	50.0
C_M^{copol}, mol %:	~ 0.3	0.8	1.5	3.1	4.6	8.6
Content, %:	15.5	37.0	16.9	20.3	6.0	4.4

The table shows that the resin produced with a supported Ziegler-Natta catalyst is a mixture of macromolecules with a very different composition. The 1-butene content in the fractions varies from very low (fractions A and B, C_M^{copol} below 1%) to very high (fraction F, C_M^{copol} ~ 9 mol %). The latter fraction has a very low crystallinity degree; it is easily dissolved at room temperature in such solvents as hexane or heptane. Commercial manufacturers measure the amount of this resin component by extracting it with a light solvent (usually hexane) and report it as *soluble material* or *extractables*. This analysis is described in Chapter 7. The presence of this copolymer fraction explains the high fracture toughness of LLDPE resins produced with Ziegler-Natta catalysts: the material has a very low crystallinity degree; it forms a separate phase and performs a function similar to that of the rubberlike second phase in high impact resins [67].

Chromium oxide catalysts are widely used for the synthesis of HDPE resins. Table 1.4 shows that these catalysts contain many different types of active centers with different kinetic parameters and produce polymer mixtures with a broad molecular weight distribution. Ethylene homopolymers (HDPE resins) produced with chromium oxide catalysts contain only linear polymer chains. Their crystallinity degree is ~70% and they melt in a narrow temperature range, 134.3 to 135 °C. CRYSTAF curves of these HDPE resins contain a single narrow crystallization peak with the maximum at ~ 84 °C.

Similar chromium oxide catalysts are also used for the manufacture of ethylene/α-olefin copolymers (LDLPE resins). The CRYSTAF curve of an LDLPE resin produced with a chromium oxide catalyst at 90 °C (ethylene/1-hexene copolymer, C_M^{copol} = 4.7 mol %) is shown in Figure 1.13. The curve is the overlap of eight crystallization peaks with maximums at temperatures ranging from 80.6 to 36 °C. These results demonstrate that different active centers in chromium oxide catalysts produce copolymer molecules of different composition [68].

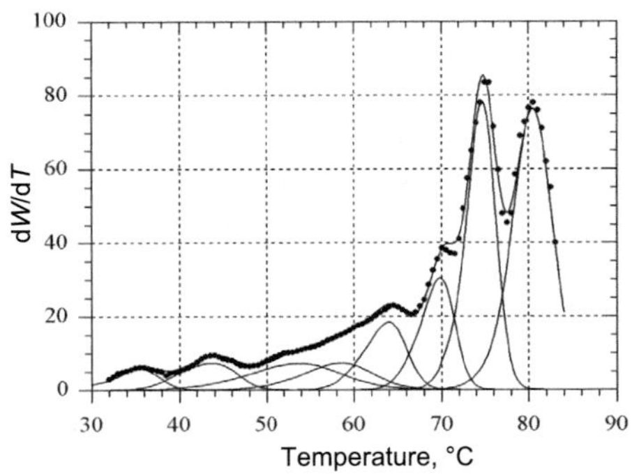

Figure 1.13
CRYSTAF curve of LDLPE resin produced with chromium oxide catalyst and its resolution into compositionally uniform components

Table 1.8 lists compositions and contents of individual fractions in this resin. The 1-hexene content in the fractions varies from ~0.7 mol % for the fraction crystallizing at the highest temperature (fraction A) to over 6 mol % for the fraction crystallizing at the lowest temperature. In addition, the resin contains 12.5% of a completely amorphous material with an even higher1-hexene content.

Table 1.8 Components of VLDPE Resin Produced with Chromium Oxide Catalyst at 90 °C (Ethylene/1-Hexene Copolymer with C_M^{copol} = 4.7 mol %); CRYSTAF Analysis

Fraction:	A	B	C	D	E	F	G	H
T_{cryst}, °C:	80.6	74.9	70.2	64.6	59.4	54.7	~45	~36
C_M^{copol}, mol %:	~0.7	1.3	1.8	2.5	3.1	3.6	~5	~6.5
Content, %:	29.7	22.7	10.0	7.4	4.9	6.5	~4	~2.5

Both Figure 1.12 and Figure 1.13 present examples of compositional nonuniformity in LLDPE resins that develops naturally, because the catalysts have several types of active centers with different copolymerization ability. Some commercial resins with strongly expressed compositional nonuniformity are produced artificially by blending two different LLDPE resins. Such blending is very difficult to achieve if one attempts to physically mix comparable amounts of two LLDPE resins with different properties. However, the blending is possible if both resins are manufactured separately as solutions in a heavy hydrocarbon solvent (Section 1.3) and then both solutions are mixed at a high temperature.

Figure 1.14 shows the CRYSTAF curve of a commercially manufactured blend of two ethylene/1-octene LLDPE resins. The first component in the blend is a resin produced with a Ziegler-Natta catalyst designed for the use in high-temperature solution processes (Section 1.2). This material is similar in terms of its composition distribution to the ethylene/1-butene copolymer produced with a supported Ziegler-Natta catalyst, which has a TREF curve shown in Figure 1.12. As with all the other LLDPE resins produced with catalysts of this type, this resin component is compositionally nonuniform. The CRYSTAF data in Figure 1.14 show that this material contains several fractions of different composition. The largest of them is the fraction with a low 1-octene content of ~ 0.5 mol % (its crystallization peak is at 81 °C). This component also contains two copolymer fractions with a higher 1-octene content, ~ 1.5 and 2.5 mol % (the broad crystallization area between 75 and 60 °C), and an amorphous fraction with a high C_M^{copol} value, which does not crystallize. The second component in the LLDPE blend in Figure 1.14 is produced with a single-site metallocene catalyst. It is compositionally uniform; its crystallization temperature is 49.1 °C. The two resin components are blended in a 1.5:1 ratio and they both have similar average molecular weights. Such blends exhibit superior mechanical properties and a very good sealing performance in packaging applications.

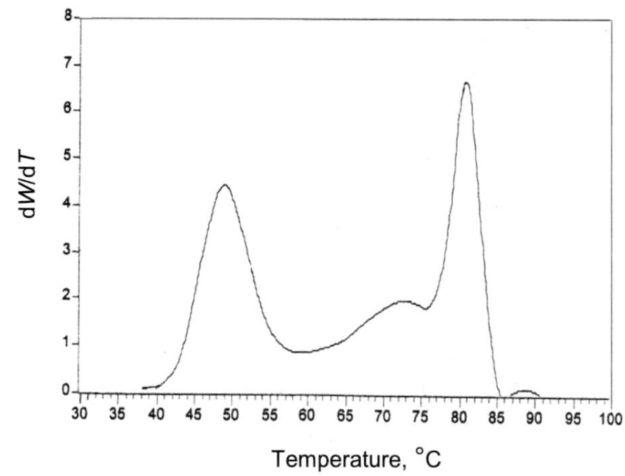

Figure 1.14 CRYSTAF curve of a blend of two LLDPE resins (ethylene/1-octene copolymers). One resin is produced with Ziegler-Natta catalyst and another with metallocene catalyst

1.9 Morphology of Polyethylene Resins

The principal crystalline form of linear polyethylene is orthorhombic [69], the same as in linear paraffins. Parameters of the orthorhombic cell are: a = 0.740 nm, b = 0.493 nm, and c (the direction of polymer chains) = 0.253 nm. This c value corresponds to the length of one ethylene unit in the polyethylene chain, $-CH_2-CH_2-$. The theoretical density of this crystalline form is 1.0 g/cm^3. The presence of short-chain branches in polyethylene chains leads to a small expansion of the orthorhombic cell; the a length increases to ~ 0.77 nm and b to ~ 0.5 nm. This expansion leads to a small decrease in crystal density.

The second important crystalline form of polyethylene is pseudo-monoclinic [69]. This crystalline modification is formed when a polyethylene resin is crystallized rapidly from the melt and when a polyethylene article is subjected to low-temperature working, for example when polyethylene film is rapidly stretched. The parameters of the pseudo-monoclinic cell are: a = 0.405 nm, b = 0.485 nm, c = 0.254 nm (the same as in the orthorhombic cell), and the angles are: $\alpha = \beta = 90°$, $\gamma = 105°$. The theoretical density of the pseudo-monoclinic form is 0.965 g/cm^3. The pseudo-monoclinic modification is stable only at temperatures below 50 °C and converts to the orthorhombic modification at higher temperatures.

All polyethylene resins, with the exception of some VLDPE resins with a very high α-olefin content (ethylene elastomers), are semi-crystalline plastics. The structure of crystalline regions in the resins depends on the level of branching in polyethylene macromolecules. Figure 1.15 shows four principal smallest morphological features in polyethylene resins of different types. A clear separation of these structural elements is possible only in the case of compositionally uniform LLDPE resins in which all the macromolecules have the same composition.

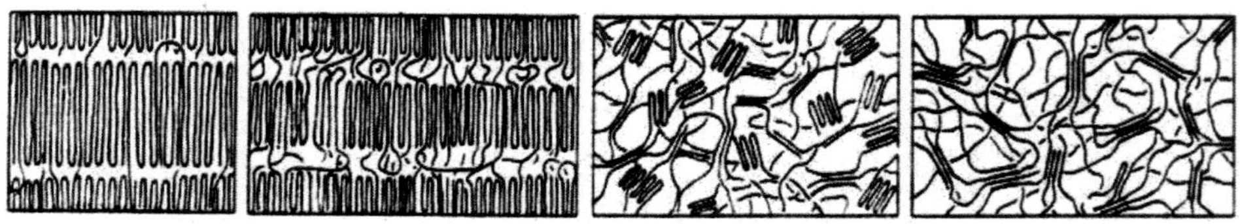

Figure 1.15 Morphological features in polyethylene resins of different types [70, 71]

The crystallization pattern in linear polyethylene has the highest regularity. The length of a typical HDPE molecule with the molecular weight of 200,000 is about 1,800 nm. Polyethylene chains are very flexible; the rotation time for the CH_2-CH_2 bond is ~40 picoseconds [72]. When polyethylene molecules crystallize, they fold on themselves numerous times and form the structural element called a *lamella*. The average length of the straight segment of a polyethylene chain in a lamella is called the lamella thickness. The distance between neighboring chains in the lamellae is 4.4 to 4.6 Å. The folds connecting adjacent straight sections of the chain consist of a few ethylene units, they form amorphous regions in HDPE resins. Each lamella has a sandwich structure; it consists of the crystalline core of folded chains and two amorphous regions at both sides of the core.

X-ray, infrared, and wide-angle neutron scattering data show that the nature of the folds depends on the crystallization conditions. When an HDPE resin is slowly crystallized from solution, the folding is tight and the chains usually fold in such a manner that it allows for the adjacent re-entry of a single chain into the same lamella [73–75]. If an HDPE resin is crystallized from solution at a constant temperature of 120 - 130 °C, the average lamella thickness is ~220 to 250 Å; it consists of 90 to 100 CH_2-CH_2 units. When the same HDPE resin is rapidly crystallized from the melt (which is the typical crystallization condition in industry), the average lamella thickness is smaller; the chain folding is much looser and resembles the old-fashion telephone switchboard, as shown in Figure 1.16 [70, 71, 73–75]. In this case, a macromolecule, after forming a loose fold, still usually enters the same lamella but some polymer chains (called *tie macromolecules*) after folding enter neighboring lamellae.

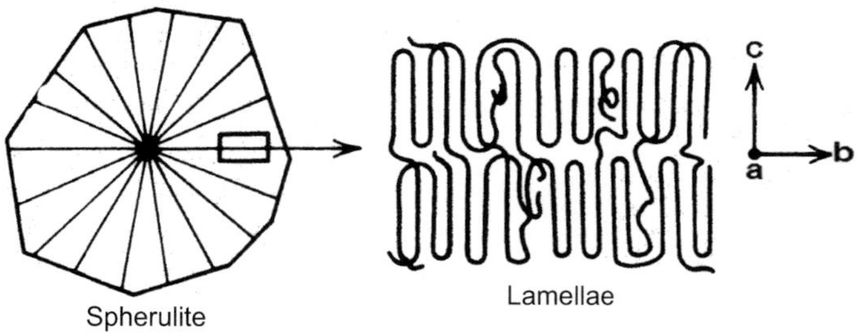

Figure 1.16 Structure of HDPE spherulite

If polyethylene macromolecules are slightly branched (for example, MDPE resins containing from 0.5 to ~1.5 mol % of α-olefin) and the fraction of long ethylene segments -M-E_n-M- is high (it is given by Equation 1.18), these segments form lamellae of the same type as in HDPE resins but their thickness is smaller. The folds in these lamellae are looser, the amorphous regions are significantly thicker, and the number of tie macromolecules is higher (Figure 1.15). A further increase in the level of branching leads to a continuing decrease in the fraction of long ethylene sequences capable of crystallization. As a result, the lamella thickness decreases even further (see Chapter 3) and the volume of the amorphous phase increases. Finally, VLDPE resins with the α-olefin content approaching 8 to 10 mol % contain only relatively short ethylene sequences and cannot form lamellae anymore.

It should be taken into account that LLDPE resins produced with Ziegler-Natta and chromium oxide catalysts are in effect mixtures of copolymer molecules with a very different α-olefin content, from 0.5 to 1.0 mol % to over 10 mol % (Section 1.8, Table 1.7 and Table 1.8). Copolymer molecules of different composition in these resins crystallize separately at different temperatures and form lamellae of different thickness (see Chapter 3).

The largest morphological unit of all HDPE and MDPE resins crystallized from the melt under typical conditions is a *spherulite*, a small anisotropic spherical object with a diameter from 1 to 5 micron. The spherulites in polyethylene film can be clearly seen under medium magnification with an optical microscope equipped with a polarizer. Figure 1.17 shows the microphotograph of typical spherulites in film made from an HDPE resin. The spherulites are formed as a result of a complex crystallization process of polyethylene macromolecules. The main structural subunits in the spherulites are thin rod-like *fibrils* (rays) radiating in all directions from the center to the periphery of the spherulite. The fibrils consist of stacked lamellae; they frequently branch and fill the whole volume of the spherulite. When polyethylene is crystallized from the melt, the *b* axis of its orthorhombic cell is the direction of chain crystallization; it is parallel to the fibril direction (Figure 1.16). The *c* axis and the *a* axis of orthorhombic crystals are perpendicular to the fibril axis.

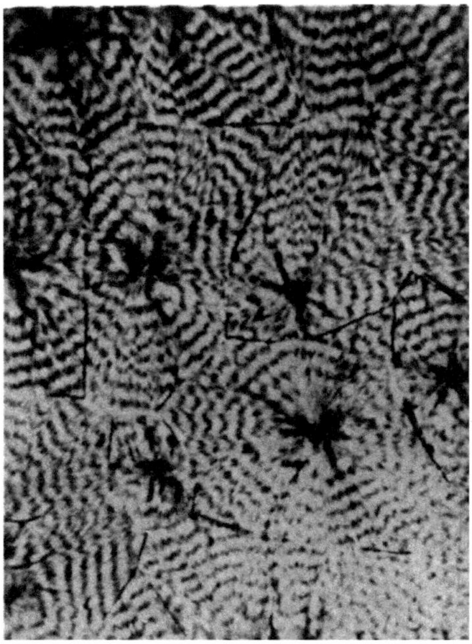

Figure 1.17
Spherulites in HDPE film [69]

1.10 Mechanical Properties of Polyethylene Resins

All grades of polyethylene resins, when used above their brittle point (typically, from −70 to −80 °C), are materials with strongly expressed ductile properties and a relatively low fracture toughness. Any forced stretching of a polyethylene sample, film or a rod, is accompanied by a series of structural transformations and mechanical changes. These mechanical changes are traditionally represented as a stress/

strain curve. Figure 1.18 shows the stress/strain curve of a typical semi-crystalline polyethylene resin, such as HPDE or LLDPE, and gives definitions of characteristic stresses σ and strains ε. The ε values in the figure are the length ratios of the strained and the original sample: ε = elongation + 1; the minimum ε value is 1. The stress/strain curve is defined by six parameters:

The *yield stress*, σ_y, and the *yield strain*, ε_y,

The *necking stress*, σ_n, and the *end-of-necking strain*, ε_n,

The *breaking* (*tensile*) *stress*, σ_{br}, and the *breaking strain*, ε_{br}.

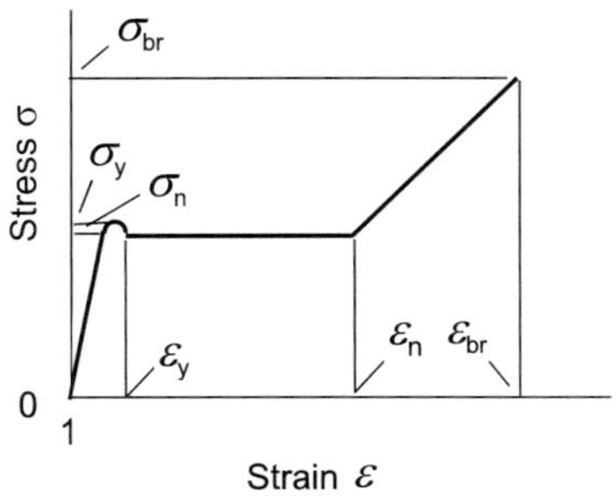

Figure 1.18
Idealized stress/strain curve of ductile polymer

All the stresses are calculated with respect to the cross section of the original sample. Typical stress values for HDPE resins are $\sigma_y \approx$ 2.9 to 3.3 kg/mm^2 (29 to 33 MPa), σ_y/σ_n, $\approx$ 1.05 to 1.1, and σ_{br} ~3 to 4 kg/mm^2 (30 to 40 MPa). The same parameters for LLDPE resins are lower: $\sigma_y \approx \sigma_n \approx$ 0.9 to 1.5 kg/mm^2 (9 to 15 MPa) depending on the resin type, $\sigma_y/\sigma_n \approx$ 1.05 to 1.1, and $\sigma_{br} \approx$ from 2.0 to 3.0 kg/mm^2 (20 to 30 MPa).

Mechanical changes shown in Figure 1.18 are accompanied by cardinal and irreversible structural changes in the polymer sample. Polyethylene spherulites deform elastically only at a deformation below ~0.5% (below the yield point). Stiffness of the sample at this stage is characterized by the elastic (Young's) modulus of the material. This parameter is defined as the tangent to the stress/strain curve ($M_{Young} \approx d(\sigma)/d(\varepsilon - 1)$ at low ε) or, more practically, as the stress at the 1% elongation point (at $\varepsilon_y - 1 = 0.01$) or at the 2% elongation point [ASTM method D5323-92 (2011)] and is called the *secant modulus*. The stress/strain curve at low elongations is practically a straight line (Figure 1.18) and the secant modulus (1%) is close to the M_{Young} value.

Further straining of the polymer sample leads to the gradual, step-by-step breaking of bridges between lamellae, slipping of lamellae in fibrils, splitting of fibrils, and other irreversible structural changes. These processes can be viewed as a succession of partial "melting" of the initial morphological features of the resins, spherulites and fibrils. At the yield point, the sample develops a neck, the area consisting of highly oriented bundles of macromolecules. As the stretching proceeds and the length of the neck increases, two of the initial morphological features, the spherulites and the fibrils, are gradually disassembled. The resin sample continues to elongate from the point ε_y to the point ε_n at a nearly constant stress σ_n until all the initial morphological structures disappear and the material becomes highly oriented, as shown in Figure 1.19.

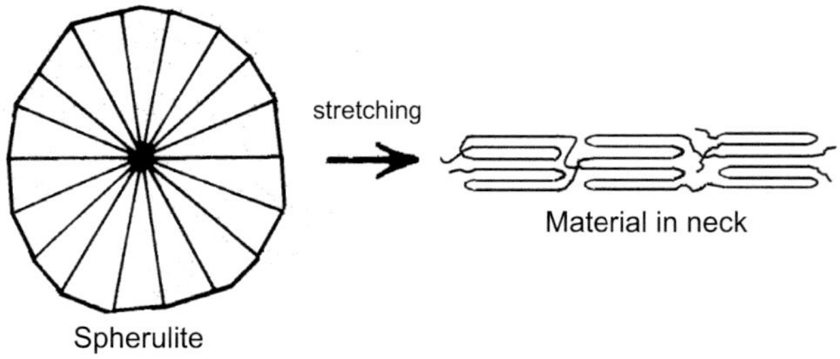

Figure 1.19 Structural changes in drawn polyethylene

After the neck region has extended to the whole length of the stretched sample, the third stage of the sample straining begins, which is called the *strain-hardening* stage. This term describes the elastic stretching of the completely oriented material; it culminates in its rupture at an arbitrary point along the stretched range of the sample. The length of the elastic deformation zone depends on the grade of the resin. Most HDPE resins break soon after the beginning of the strain-hardening stage and their σ_y and σ_{br} values are relatively close. In contrast, LLDPE resins have a relatively long strain-hardening region and their σ_{br} value is always much higher than the σ_y value. These differences are attributed to the mechanism of strain-hardening. The strain-hardening phenomenon is essentially the straining process of the segments of tie macromolecules holding together highly oriented crystalline lamellae (Figure 1.19). The number of tie molecules is significantly lower than the total number of polymer segments in the amorphous regions of stretched samples. In essence, the behavior of the tie molecules under strain is described by the stretching mechanism of amorphous elastic polymers above their glass point [76].

It should be kept in mind that all the σ values in Figure 1.18 are given with respect to the original cross section of a sample before the beginning of its deformation. The real cross section is dramatically reduced at the beginning of the necking stage (by a factor of $\sim\varepsilon_{br}^2$) and it is further reduced at the breaking point (by a factor of $\sim\varepsilon_{br}^2$); consequently, the real σ_n and σ_{br} values calculated with respect to the real cross sections of the necked and the strained sample are, respectively, much higher.

The values of all σ and ε parameters on the stress/strain curve in Figure 1.18 strongly depend on the molecular weight of polyethylene resins and their compositional uniformity. Low molecular weight HDPE resins are brittle and break at a strain of $<10\%$ without the neck development. HDPE resins with a molecular weight in the range of 80,000 to ~1,000,000 always develop a neck. Increasing the molecular weight of such resins is accompanied by a drastic decrease of the ε_{br} value, from ~15 to ~3, and by a significant increase of the σ_{br} value, from ~3 to ~6 kg/mm^2 (30 to 60 MPa).

The high fracture toughness of LLDPE resins produced with Ziegler-Natta catalysts is explained by their compositional nonuniformity (Table 1.4) [67]. These resins have two different crystalline phases. The first, dominant phase is formed by copolymer molecules with a low α-olefin content. This copolymer component is responsible for high σ_y and σ_n values and for high stiffness of the LLDPE resins (high Young's modulus). The second, minor phase is formed by copolymer molecules with a high α-olefin content. This material fills voids between crystalline lamellae (Figure 1.15) and between adjacent spherulites (Figure 1.17). This phase performs a function similar to that of the rubberlike phase in high-impact resins. The presence of this phase explains high ductile properties of the materials (high ε_n value) and their high strain-hardening parameters (high ε_{br} and σ_{br} values).

HDPE resins of ultrahigh molecular weight, one to two million (UHMW HDPE), represent a special grade of polyethylene resins (Table 1.1). Although these polymers are strictly linear, the resins have a low crystallinity degree, 30 to 40%, and low density. When stretched, these resins do not develop the neck but uniformly elongate by 200 to 400%. UHMW HDPE resins can be processed at pressures of 2,000 to 3,000 atm and temperature below 100 °C into filaments or film that are very strongly stretched and nearly perfectly oriented, both in the crystalline and in the amorphous phase. These items contain a large fraction of tie macromolecules [77, 78]; they have a very high tensile strength (σ_{br} = 50 to 60 kg/mm^2 or 500 to 600 MPa) and a very high Young's modulus, up to ~$1 \cdot 10^4$ kg/mm^2 (~$1 \cdot 10^5$ MPa).

1.11 Recycling of Polyethylene Resins

At the present time, HDPE resins are one of the two general-purpose plastics commercially recycled on a large scale in the United States and in Europe (the other such plastic being polyethylene terephthalate, PET). The recycling code for commercial HDPE products (resin ID) is "2". HDPE resins are highly durable non-biodegradable materials; they constitute over 35% of all post-consumer recycling streams. HDPE bottles and industrial household containers (mostly detergent vessels) are recycled at the highest rate, ~30%.

Potential advantages of HDPE recycling include several economic and environmental factors. It is usually cheaper to make HDPE articles from recycled matter than to use new materials. Recycling these products takes about half as much energy as their incineration; it reduces expenses associated with manufacturing products made from HDPE resins and reduces the expense of waste disposal (landfill costs). Recycling significantly reduces the waste volume. HDPE containers have a very low weight/volume ratio; they are sturdy and degrade in landfills very slowly. The principal disadvantages of HDPE recycling are two: low cost of virgin resins and complexity and high cost of sorting the recycle streams.

The process of HDPE recycling consists of two stages: (a) recovery and collection of scrap or waste plastic and (b) reprocessing the material into useful products. In turn, the reprocessing stage involves several steps: sorting the combined collected recycle stream (a mixture of HDPE and PET articles), shredding, washing, pelletization, and the manufacture of new HDPE items. Intensely or dark colored plastic is usually discarded during the sorting step because it is difficult to reuse. The waste-separating facilities typically use float-sink tanks to separate shredded polymers and to remove contaminants. In general, the cost-effective and efficient separation of mixed plastic streams and their cleaning from hard-to-remove residues is the biggest challenge facing the recycling industry. The quality of any plastic degrades once it has been used and recycled HDPE resins are generally downgraded into lower-quality materials of lesser value. For the same reasons, reprocessed HDPE is not used for food or pharmaceutical packaging.

Recycled HDPE is converted into a wide variety of products. Products commonly manufactured from recycled HDPE include the following:

- Building materials. They can be mixed with fiberglass strand for added strength and structural integrity. These materials are formed into solid and hollow profiles or into injection molded parts.
- Garden furniture (tables, benches)
- Trash cans and crates
- Envelopes and separators for shipping large-size items
- Paneling
- Water-drain pipe
- Wood/plastic composites (plastic lumber)

Wood/plastic composites are produced by thoroughly mixing ground wood particles (waste products of the wood industry) with recycled HDPE and extruding the material into the desired shape. These composites are widely used for construction of outdoor deck floors, railings, fences, landscaping timber, siding, park benches, and window and door frames. Their principal advantage over wood is the ability of the material to be molded into almost any desired shape. Although the composites contain over 50% of finely dispersed cellulose, their mechanical behavior is similar to that of neat HDPE. The composites do not corrode and are highly resistant to rot, decay, and attack by marine borers. Unlike wood timber or concrete, they too can be recycled into a new wood/plastic composite.

References

1. Nowlin, T.E., Mink, R.I., Kissin, Y.V., In *Transition Metal Polymerization Catalysts.* Hohn, R., Mathers, R.T. (Eds.) (2009) John Wiley & Sons, Hoboken, p. 131
2. Krentsel, B.A., Kissin, Y.V., Kleiner, V.I., Stotskaya, S.S., *Polymers and Copolymers of Higher α-Olefins* (1997) Hanser, Munich, Chapter 8
3. Kissin, Y.V., *Alkene Polymerization Reactions with Transition Metal Catalysts* (2008) Elsevier, Amsterdam, Chapters 3 and 4
4. Kissin, Y.V., In *Handbook of Industrial Polyethylene and Technology,* Part 1, Spalding M.A., Chatterjee, A.M. (Eds.) (2017) Scrivener Publishing and Wiley: Beverly, MA, p. 25
5. *AkzoNobel Polymer Chemicals, Initiators and Reactor Additives for Thermoplastics* (2014)
6. Berbee, O.J., Den, D.C., Hosman, C.J., Karjala, T.P., Hinrichs, S., *World Patent Appls.* WO 2012 044 504 A1 (2012), WO 2013 078 224 A1 (2013)
7. Bovey, F.A., Schilling, F.C., McCrackin, F.L., Wagner, H.L., *Macromolecules* (1976) 9, p. 76
8. Kulin, L.I., Meijerink, N.L., Starck, P., *Pure Appl. Chem.* (1988) 60, p. 1403
9. Kissin Y.V., Nowlin T.E., Mink R.I., In *Handbook of Transition Metal Polymerization Catalysts,* 2nd Ed., R. Hoff (Ed.) Wiley: New York, (2018), p. 189
10. Nowlin, T.E., *Business and Technology of the Global Polyethylene Industry,* Scrivener Publishing LLC, New York (2014)
11. Makio, H., Terao, H., Iwashita, A., Fujita, T., *Chem. Rev.* (2011) 111, p. 2363
12. Nakata, N., Toda, T., Ishii, A., *Polym. Chem.* (2011) 2, p. 1597
13. Nagai, N., Mitsuzuka, M., Nakai, K., Isokawa, M., Nakatsuka, S., Taneichi, D., Honma, S., Narutaki, T., *World Patent Appl.* WO 2005 073 282 (2005)
14. Muruganandam, N., Abichandani, J., Terry, K.A., Patel, H.G., Rodriguez, G., *U.S. Patent* 8,088,871 (2012)
15. Moritz, M.S., Baier, C., Zuideveld, M.A., Mecking, S., *Angew. Chem. Int. Ed.* (2014) 53, p. 9722
16. *Metallocene Technology in Commercial Applications.* Benedikt, G.M. (Ed.) (1999) Plastics Design Library, Norwich
17. *Polyolefins Through the 80s – A Time of Change* (1983) SRI International, Menlo Park, CA
18. Choi, K.-Y., Ray, W.H., *J. Macromol. Sci.-Rev. Macromol. Chem., Phys.* (1985) C25, p. 57
19. *Advances in Polyolefins.* Cheng, T.C., Seymour, R.B. (Eds.) (1989) Plenum, New York
20. Karol, F.J., Jacobson, F.I., In *Catalytic Polymerization of Olefins.* Keii, T., Soga, K. (Eds.) (1986) Elsevier, Amsterdam, p. 323
21. Hsieh, J.T.T., Simondsen, J.C., *U.S. Patent* 5 096 868 (1992)
22. Lo, F.Y., Nowlin, T.E., Shirodkar, P.P., *U.S. Patent* 5 032 562 (1991)
23. Miller, A.R., *U.S. Patent* 4 003 712 (1977)
24. Goeke, G.L., Wagner, B.E., Karol, F.J., *U.S. Patent* 4 302 565 (1981)
25. Furtek, A.B., Proceedings from *MetCon'93,* Houston, TX, (1993) p. 125
26. Ali, A.H., Hagerty, R.O., Ong, S.C., *Canadian Patent* 2 078 655 (1993)
27. Yi, K.C.H., Maraschin, N.J., In *Maack'90 International Business Conference,* (1990) Zurich
28. Miller, B., *Plastics World* (1992) p. 46
29. Knuuttila, H., Lehtinen, A., Nummila-Pakarinen, A., *Adv. Polym. Sci.* (2004) 169, p. 13
30. Short, J.N., In *Transition Metal Catalyzed Polymerizations: Alkenes and Dienes.* Quirk, R.P. (Ed.) (1983) Harwood Academic Publishers, New York, p. 651
31. Hamilton, M.A., Zboril, V.G., *U.S. Patent* 4 547 473 (1985)

32. Lai, S.Y., Wilson, J.R., Knight, G.W., Stevens, J.C., Chum, P.S., *U.S. Patents* 5 272 236 (1993) and 5 278 272 (1994)
33. McKnight, A.L., Waymouth, R.M., *Macromolecules* (1999) 32, p. 2816
34. Shapiro, P.J., Bunel, E.E., Schaefer, W.P., Bercaw, J.E., *Organometallics* (1990) 9, p. 867
35. Gates, D.P., Svejda, S.A., Oñate, E., Killian, C.M., Johnson, L.K., White, P.S., Brookhart, M., *Macromolecules* (2000) 33, p. 2320
36. Schultz, C.S., DeSimone, J.M., Brookhart, M., *Organometallics* (2001) 20, p. 16
37. Flory, P.J., *Principles of Polymer Chemistry* (1953) Cornell University Press, Ithaca
38. Schulz, G.V., *Z. Phys. Chem., B* (1939) 43, p. 25
39. Bamford, C.H., In *Encyclopedia of Polymer Science and Engineering.* Mark, H.F., Bikales, N.M., Overberger, C.G., Manges, G. (Eds.) (1986) John Wiley & Sons, New York, Vol. 13, p. 845
40. Peebles, L.H., *Molecular Weight Distributions in Polymers* (1971) Wiley-Interscience, New York
41. You, W., Kirkland, J., Bly, D., *Modern Size-Exclusion Liquid Chromatography* (1979) John Wiley & Sons, New York, Chapter 3
42. Kissin, Y.V., *J. Polym. Sci., Part A: Polym. Chem.* (1995) 33, p. 227
43. Soares, J.B.P., Hamielec, A.E., *Polymer* (1995) 36, p. 2257
44. Faldi, A., Soares, J.B.P., *Polymer* (2001) 42, p. 3057
45. Ver Strate, G., Cozewith, C., West, R.K., Davis, W.M., Capone, G.A., *Macromolecules* (1999) 32, p. 3837
46. Luruli, N., Grumel, V., Brüll, R., Du Toit, A., Pasch, H., Van Reenen, A.J., Raubenheimer, H.G., *J. Polym. Sci., Part A: Polym. Chem.* (2004) 42, p. 5121
47. DesLauriers, P.J., ACS Symp. Series (2006) 893, Chapter 13, p. 210
48. Thorshaug, K., Mendici, R., Boggioni, L., Tritto, I., Trinkle, S., Friedrich, C., Mülhaupt, R., *Macromolecules* (2002) 35, p. 2903
49. Kissin, Y.V., Mink, R.I., *J. Polym. Sci., Part A: Polym. Chem.* (2010) 48, p. 4219
50. Zhang, L.-T., Fan, Z.Q., Fu, Z.-S., *Chinese J. Polym. Sci.* (2008) 26, p. 605
51. Kissin, Y.V., Nowlin, T.E., Mink, R.I., Brandolini, A.J., In *Metalorganic Catalysts for Synthesis and Polymerization.* Kaminsky, W., ed. (1999) Springer, Berlin, 1999, p. 60
52. Kissin, Y.V., Nowlin, T.E., Mink, R.I., Brandolini, A.J., *Topics Catal.* (1999) 7, p. 69
53. Kissin, Y.V., *Adv. Polym. Sci.* (1974) 15, p. 91
54. Soares, J.B.P., Hamielec, A.E., *Polymer* (1995) 36, p. 1639
55. Mingozzi, I., Cecchin, G., Morini, G., *Int. J. Polym. Anal. Charact.* (1997) 3, p. 293
56. Mirabella, F.M., *J. Polym. Sci., Part B: Polym. Phys.* (2001) 39, pp. 2800, 2819
57. Kissin, Y.V., Mirabella, F.M., Meverden, C.C., *J. Polym. Sci., Part A: Polym. Chem.* (2005) 43, p. 4351
58. Wild, L., *Adv. Polym. Sci.* (1991) 98, p. 1
59. Kakugo, M., Miyatake, T., Naito, Y., Mizunuma, K., In *Transition Metal Catalyzed Polymerizations. Ziegler-Natta and Metathesis Polymerization.* Quirk, R.P. (Ed.) (1989) Cambridge University Press, New York, p. 624
60. Monrabal, B., *J. Appl. Polym. Sci.* (1994) 52, p. 491
61. Soares, J.B.P., Monrabal, B., *Macromol. Chem. Phys.* (1998) 199, p. 1917
62. Soares, J.B.P., Hamiliec, A.E., *Macromol. Theory Simul.* (1995) 4, p. 305
63. Monrabal, B., *Makromol. Chem., Macromol. Symp.* (1996) 110, p. 81
64. Britto, L.J.D., Soares, J.B.P., Penlidis, A., *J. Polym. Sci., Part B: Polym. Phys.* (1999) 37, p. 539
65. Chu, K.J., Soares, J.B.P., Penlidis, A., *Macromol. Chem. Phys.* (2000) 201, p. 340

66. Kissin, Y. V., Fruitwala, H. A., *J. Appl. Polym. Sci.* (2007) 106, p. 3872
67. Mirabella, F. M., Westphal, S. P., Fernando, P. L., Ford, E. A., Williams, J. G., *J. Polym. Sci., Part B: Polym. Phys.* (1988) 26, p. 1995
68. Kissin, Y. V., Brandolini, A. J., Garlick, J. L., *J. Polym. Sci., Part A: Polym. Chem.* (2008) 46, p. 5315
69. Miller, R. L., In *Crystalline Olefin Polymers, Part 1.* Ruff, R. A. V., Doak, K. W. (Eds.) (1965) Interscience, New York, p. 577
70. Stamm, M., *J. Polym. Sci., Part B: Polym. Phys.* (1982) 20, p. 235
71. Wignall, G. D., Mandelkern, L., Edwards, C., Glotin, M., *J. Polym. Sci., Part B: Polym. Phys.* (1982) 20, p. 245
72. Zheng, J., Kwak, K., Xie, J., Fayer, M. D., *Science* (2006) 313, p. 1951
73. Jing, X., Krimm, S., *J. Polym. Sci., Polym. Lett. Ed.* (1983) 21, p. 123
74. Ching, J. H. C., Krimm, S., *Mater. Sci. Technol.* (2008) 46, p. 4181
75. Tashiro, K., *Acta Polym.* (2003) 46, p. 100
76. Ward, I. M., *Mechanical Properties of Solid Polymers*, 2nd Ed., John Wiley & Sons, New York, (1983), Chapter 3
77. Ward, I. M., *Polym. Eng. Sci.* (1984) 24, p. 724
78. Zachariades, A. E., Porter, R. S., *J. Macromol. Sci., Part B: Phys.* (1981) 19, p. 377

2 Melt Index and Melt Flow Ratio of Polyethylene Resin

2.1 Introduction

The melt index of a polyethylene resin is universally accepted in the industry as an indicator of the resin's average molecular weight. Melt indexes are routinely measured both in the industrial environment and in laboratory using simple automated equipment and simple standard procedures (Section 2.1.1). The polymer science operates with a different set of parameters describing molecular weight characteristics of ethylene polymers. The three most often used parameters are the weight-average molecular weight M_w, the number-average molecular weight M_n, and the width of the molecular weight distribution, the M_w/M_n ratio. Measurement methods and the statistical meaning of these parameters are described in Section 1.5 of Chapter 1.

This chapter describes correlations between these two sets of molecular weight characteristics.

2.1.1 Measurement Method of Melt Index

Melt indexes of polyethylene resins are measured using the instrument called an *extrusion plastometer*. The measurement procedures are specified in the American Society for Testing and Materials (ASTM) method D1238-10, Conditions E and F, and in the International Organization for Standardization (ISO) Method 1133. Figure 2.1 (a) shows the schematics of the melt index measurement. A small amount of resin (~6 gram) is placed inside the heated cylindrical barrel (9.55 mm diameter) with a metal die at its bottom. The die has a round capillary opening, 2.095 mm in diameter and 8 mm in length. Some procedures use a shorter die (a half-die) 4 mm long with a 1.0475 mm capillary opening. A metal plunger with a long narrow stem is placed on the top of the resin sample. The resin is kept at 190 °C for 6 minutes to achieve complete melting. After that, the melt is pressurized by placing a cylindrical metal weight on the upper tip of the stem. The pressure of the plunger forces the melt through the capillary opening. By definition, the melt index of a polyethylene resin is the weight of the melt extruded through the capillary over a period of 10 minutes, that is, the melt index is a measure of the melt flow rate (g/10 min). Depending on the type of polyethylene resin, four different standard weights (the total of the cylindrical metal weight, the plunger, and the rod) are used to measure melt indexes, 2.16, 5.16, 10.16, and 21.6 kg. The melt indexes are respectively designated as I_2, I_5, I_{10}, and I_{21}. If no indication of the weight is given, the I_2 value is traditionally reported. In the earlier designs of extrusion plastometers, the polymer extrudate was collected, cooled, and its amount was determined by weighing. Modern extrusion plastometers use an automated procedure; they measure the speed of the downward movement of the plunger under a given weight.

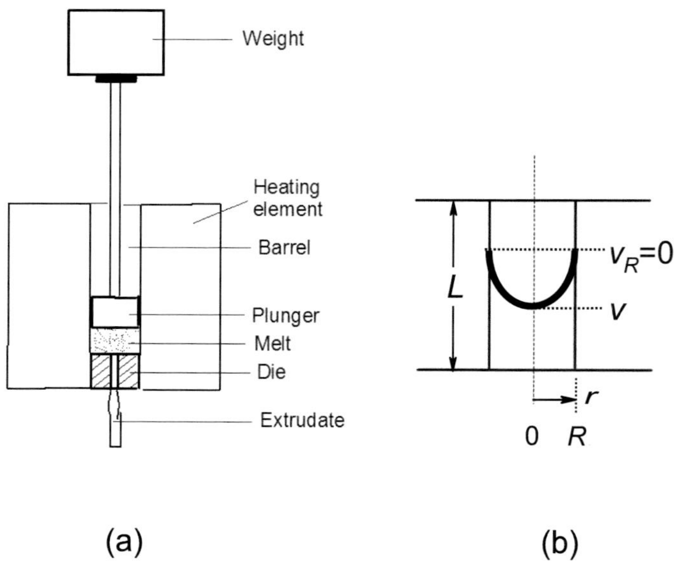

Figure 2.1 (a) Schematics of melt index measurement; (b) flow of viscous liquid through capillary

2.1.2 Empirical Correlations between Melt Index and Molecular Weight

If one is dealing with polyethylene resins of a single particular type (for example, resins produced with the same catalyst under different reactor conditions), the melt index is a precise relative measure of the resin's weight-average molecular weight M_w. Several empirical correlations between M_w values and melt indexes of various polyethylene resins were developed. Figure 2.2 shows one such example for a series of resins with a narrow molecular weight distribution (M_w/M_n ~ 2) produced with a soluble metallocene catalyst.

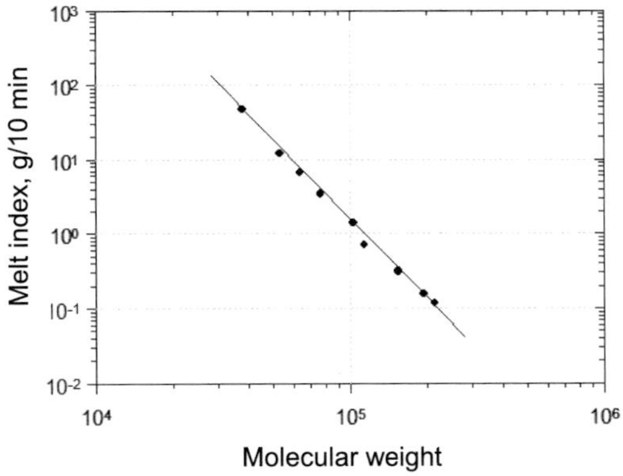

Figure 2.2
Dependence between the melt index I_2 and the average molecular weight M_w for polyethylene resins with narrow molecular weight distribution

In general, such dependencies are represented by empirical correlations of the general type: melt index = $K \cdot M_w^{-\alpha}$ or log(melt index) = log(K) − α·log(M_w) with the α value between 3.4 and 3.7 [1 - 6]. For example, the data in Figure 2.2 can be represented by the expression:

$$\log(I_2) = 17.45 - 3.451 \cdot \log(M_w) \quad (2.1)$$

Another empirical correlation for the same type of resins is:

$$I_2 = 6.16 \cdot 10^{17} \cdot M_w^{-3.5} \quad (2.2)$$

Several examples of the latter dependence are given in Table 2.1.

Table 2.1 Correlation between Melt Index and Molecular Weight of Polyethylene Resins with Narrow Molecular Weight Distribution

M_w:	90,000	92,500	95,000	97,500	100,000	150,000
I_2, g/10 min:	2.24	2.04	1.85	1.69	1.55	0.37

Modern extrusion plastometers measure melt indexes with a precision of ± 2 to 3%. The data in Table 2.1 show that a 2.5% difference between M_w values of two resins (a difference which would require the use of very precise and expensive equipment and would take several hours to determine) can be easily noticed when the resins' I_2 values are compared. In this example, the I_2 values differ by 8.3%. Such a difference can be easily established in routine I_2 measurements over a period of ~ 15 minutes. Similarly, a 50% difference in the molecular weight leads to a fourfold change in the I_2 value.

The principal uncertainty in using dependencies like those in Equation 2.1 and Equation 2.2 is that numerical parameters K and α in such expressions depend on the type of the resin and, first of all, on its molecular weight distribution [1–6]. For example, the correlation between I_2 and M_w for polymers with a broader molecular weight distribution, HDPE resins produced with supported Ziegler-Natta catalysts, uses different K and α parameters [7, 8]:

$$I_2 = 9.52 \cdot 10^{18} \cdot M_w^{-3.79} \tag{2.3}$$

The molecular weight distribution of a polyethylene resin is characterized in the industry by the ratio of two melt indexes measured with different loads. The commonly used ratios are the I_{21}/I_2 ratio (it is often called the *melt flow ratio*, or *MFR*) and the I_{10}/I_2 ratio. Both these ratios approximately correlate with the M_w/M_n value of a resin. For example, an empirical correlation between the M_w/M_n ratio and the I_{21}/I_2 ratio for LLDPE resins with a medium-broad molecular weight distribution produced with supported Ziegler-Natta catalysts is [7, 8]:

$$M_w/M_n \approx 0.238 \cdot (I_{21}/I_2) - 2.4 \tag{2.4}$$

Melt index ratios of polyethylene resins are quite sensitive to the presence of long-chain branches in polymer molecules. As an example, if a compositionally uniform ethylene/α-olefin copolymer produced with a metallocene catalyst has a small amount of long-chain branches, its I_{10}/I_2 ratio increases from ~5 to a range of 8 to 12 [9–12].

2.2 Basics of Polymer Rheology. Melt Flow through a Capillary

All polyethylene resins in the molten state are viscous non-Newtonian liquids. This statement requires a definition of what Newtonian and non-Newtonian liquids are. In very simple terms, a Newtonian liquid is a liquid in which the viscosity does not depend on the shear (the speed of the melt flow) applied to it. Because any shear would force the liquid to flow, a Newtonian liquid can be also defined as a liquid in which the viscosity does not depend on the speed at which it flows. The melt of a polyethylene resin never exhibits such a behavior; in contrast, the faster it flows the less viscous it becomes. This phenomenon becomes immediately obvious when melt indexes of the resins are measured under different loads. If a polyethylene resin were a Newtonian liquid, its I_{21}/I_2 value would always be ~10, whatever the molecular weight of the resin is (Section 2.2.1). In reality, the I_{21}/I_2 ratio of any polyethylene melt is always higher than 10. When the load is increased from 2.16

to 21.6 kg, the melt starts flowing through a capillary much faster, its effective melt viscosity decreases, and the amount of the melt discharged through a capillary increases disproportionately. The lowest experimental value of the melt flow ratio I_{21}/I_2, from 15 to 16, was measured for polyethylene resins with a narrow molecular weight distribution (M_w/M_n ratio ~ 2) prepared with soluble single-site metallocene catalysts, whereas resins with a very broad or bimodal molecular weight distribution (see Figure 1.8 and Figure 1.9) can have the I_{21}/I_2 ratio in excess of 150.

This deviation from the Newtonian flow behavior has a simple explanation. Every macromolecule with a given molecular weight (*MW*), whether it is in solution or in the melt, acquires the form of a sphere (a coil) with the radius (radius of gyration) R_g:

$$R_g = k_g \cdot (0.5/6^{0.5}) \cdot MW^{0.5} \tag{2.5}$$

where k_g is a parameter characterizing the "quality" of the medium surrounding the coil. The "better" the solvent quality (such solvents are called "good"), the higher is the k_g value and the bigger is the radius of the coil.

When an external stress is applied to the coil, its shape changes from spherical to ellipsoidal (Figure 2.3) and the higher the stress (the flow velocity), the more elongated the ellipsoid becomes. Because the resistance of a polymer coil to movement in a flow is proportional to its cross section in the direction of the flow (to the square of its effective radius shown in Figure 2.3), any decrease of the effective radius reduces the coil's resistance to flow and, therefore, increases the overall speed of the flowing melt.

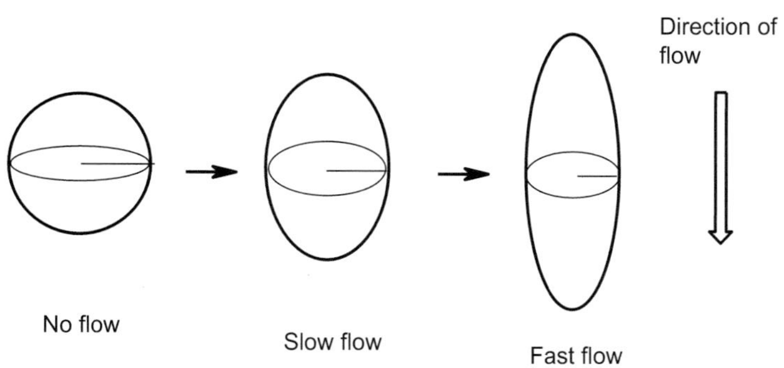

Figure 2.3 Distortion of macromolecular coil under stress

When the shear (the speed of the melt flow) is very small, viscous properties of the melt are quite similar to those of a Newtonian liquid. The viscosity η of a polymer melt measured at a very low shear is called the zero-shear viscosity, η_0. This value is closely related to the average molecular weight of the resin. If all macromolecules in a given polymer sample have the same length (such materials can be synthesized under special conditions, see Section 2.3), the correlation between the zero-shear viscosity and the polymerization degree n is described by the firmly established theoretical expression, which has been confirmed in numerous experimental measurements for polyethylene fractions with a narrow molecular weight distribution [12–14]:

$$\eta_{0,n} = k \cdot n^{\gamma} \tag{2.6}$$

with the parameter $\gamma = 3.5$. The k parameter in Equation 2.6 is not known in advance; it should be estimated from experimental data, as described below.

The issue of interpreting the physical meaning of the melt index can be separated into three subjects:

1. What is the melt index of a viscous Newtonian liquid? This subject is discussed in Section 2.2.1 and Section 2.2.2.
2. What is the melt index of the simplest non-Newtonian liquid, a polymer with all macromolecules of the same polymerization degree? This subject is discussed in Section 2.3.
3. What is the melt index of a real polymer containing macromolecules of many different lengths? This subject is discussed in Section 2.4.

2.2.1 Flow of Polymer Melt through a Cylindrical Capillary

Figure 2.1(b) shows the diagram of a liquid flow through a cylindrical capillary with the radius R and the length L under the pressure P. In the case of extrusion plastometers used for the melt index measurement, P is calculated as:

$$P(\mathrm{MPa}) = 9.81 \cdot Weight(\mathrm{kg}) \Big/ \left[\pi \left(D_{\mathrm{bar}} \cdot 10^{-3}/2\right)^2\right] \cdot 10^{-6} \tag{2.7}$$

where D_{bar} is the diameter of the barrel in the plastometer, 9.55 mm (Figure 2.1(a)) and *Weight* is the weight of the load.

When uncompressible viscous liquid flows through a capillary, the flow remains laminar in the total range of the flow speeds. The velocity v_r of the laminar flow depends on the distance r from the axis of the capillary. The velocity at the capillary wall is zero ($v_R = 0$ at $r = R$) and it reaches a certain maximum value along the axis, at $r = 0$.

The following definitions are commonly used in polymer rheology for the description of a polymer flow through a capillary [9, 15]:

1. The definition of the shear stress D_r at a radius r as a gradient of the linear melt flow velocity v_r:

$$D_r = -dv_r/dr \tag{2.8}$$

2. The definition of the shear strain τ_r in the melt flowing through a capillary with the radius R and the length L (the Stokes equation):

$$\tau_r = (P/2L)\cdot r \tag{2.9}$$

3. The shear strain is equal to zero in the center of the capillary (at $r = 0$) and it reaches a maximum at the capillary wall, at $r = R$.
4. The definition of the melt viscosity η:

$$\eta = \tau_r/D_r \tag{2.10}$$

5. The principal expression for the viscous flow of a polymer melt through a capillary (a combination of Equation 2.9 and Equation 2.10):

$$(P/2L)\cdot r(=\tau_r) = \eta \cdot D_r \tag{2.11}$$

   This expression represents the starting point for all calculations of a non-Newtonian flow.
6. The volume flow rate dV_r in an elementary layer at a radius r from the capillary axis:

$$dV_r = 2\pi \cdot r \cdot dr \cdot v_r \tag{2.12}$$

   where dr is the thickness of the elementary layer and $2\pi r \cdot dr$ is its cross-section. Boundary conditions at the wall of the capillary (at $r = R$) are: $\tau_R = (P/2L)\cdot R$ and $\eta_R = \tau_R/D_r$. Hence, $D_r = \tau_R/\eta_R = (P/2L)\cdot R/\eta_R$, which signifies that the shear stress D_r in the elementary layer is the highest at the capillary wall.
7. The total volume flow rate V through the capillary is the integral of dV_r over r:

$$V = \int_0^R 2\pi \cdot r \cdot v_r \cdot dr \tag{2.13}$$

The traditional approach to solving Equation 2.13 analytically involves a change of the variable from the radius r to the shear strain τ_r [12]. The following replacements are made: $dr = (P/2L)^{-1}\cdot d\tau_r$ (from Equation 2.9) and $dv_r = -(\tau_r/\eta_r)\cdot dr = (P/2L)^{-1}\cdot \eta_r^{-1}\cdot \tau_r \cdot d\tau$ (from Equation 2.8, Equation 2.9, and Equation 2.10). These replacements lead to the formulation of a new expression for the linear flow velocity v_r at a distance r from the capillary axis:

$$v_r = \int_{v_{r=0}}^{v_r} dv_r = v_{r=0} - (P/2L)^{-1}\cdot \int_0^{(P/2L)\cdot r} \eta_r^{-1}\cdot \tau_r \cdot d\tau \tag{2.14}$$

The lower limit in the first integral, $v_{r=0}$, is the maximum melt velocity in the center of the capillary.

Potentially, the integration of Equation 2.14 from 0 to $\tau_r = (P/2L) \cdot r$ should give the expression for the linear flow velocity v_r at a distance r. If this expression is introduced into Equation 2.13, it allows the calculation of the total melt flow rate V through a capillary. Indeed, this solution is well known for Newtonian liquids (Section 2.2.2). Unfortunately, this general approach meets the principal difficulty in the case of non-Newtonian liquids: the term η_r in Equation 2.14 is not a parameter (a constant value) but a variable; the melt viscosity of any non-Newtonian liquid depends on the shear stress D_r. Although general analytical solutions of this problem are not possible, Equation 2.14 and Equation 2.13 can be solved numerically, as described in Section 2.4.2.

2.2.2 Melt Index of Newtonian Melt

The analytical solution of Equation 2.14 is known only for a Newtonian liquid when η_r is a constant value, the zero-shear viscosity η_0 [12]. In this case, integration of Equation 2.14 gives the following expression for the linear flow velocity v_r at a distance r from the capillary axis:

$$v_r = v_{r=0} - 0.5\left(P/2L\right)^{-1} \cdot \eta_0^{-1} \cdot \tau_r^2 = v_{r=0} - 0.5\left(P/2L\right) \cdot \eta_0^{-1} \cdot r^2 \tag{2.15}$$

The $v_{r=0}$ value is determined from the boundary condition: the melt does not slip at the capillary wall, that is, $v_r = 0$ at $r = R$; therefore, $v_{r=0} = 0.5 \cdot (P/2L) \cdot \eta_0^{-1} \cdot R^2$. Under this condition, the final expression for the velocity v_r of a Newtonian liquid as a function of the radius r is the parabolic function (see Figure 2.1(b)):

$$v_r = 0.5\left(P/2L\right) \cdot \eta_0^{-1} \cdot (R^2 - r^2) \tag{2.16}$$

Introduction of Equation 2.16 into Equation 2.13 and integration from $r = 0$ to R gives the total melt flow rate of a Newtonian liquid through a capillary:

$$V^{\text{Newton}} = \pi \cdot P \cdot R^4 / (8 \cdot L \cdot \eta_0) \tag{2.17}$$

Equation 2.17 (the Hagen-Poiseuille equation) states that when a Newtonian liquid with the zero-shear viscosity η_0 flows through a capillary, the volume rate is proportional to the pressure P in the capillary; that is, it is proportional to the weight of the load placed on the stem of the plunger in an extrusion plastometer (see Figure 2.1(a)). The Hagen-Poiseuille equation also states that the volume rate of the outflowing Newtonian liquid is proportional to the square of the capillary cross section ($S_{\text{capillary}} = \pi \cdot R^2$) and that it is in a reciprocal dependence on the capillary length L and the zero-shear viscosity of the liquid.

2.3 Melt Flow of Monodisperse Polyethylene Resins

All polyethylene resins are non-Newtonian liquids, and Equation 2.17 does not describe their flow through a capillary because the viscosity of all these resins is not a constant value but varies depending on the velocity of the flow, v_r. The simplest example of such a deviation can be observed experimentally when one measures the melt flow of model polymers that have all their macromolecules the same length (the same polymerization degree n). Such polymers are called monodisperse polymers. Polyethylene resins of this type cannot be produced from ethylene with any polymerization catalyst. However, they have been prepared artificially in a series of two chemical reactions that include anionic polymerization of butadiene with a butyl lithium initiator followed by hydrogenation of the produced polybutadiene [16]. The final products of these two reactions are very similar to ethylene/1-butene copolymers in terms of their molecular structure but they have a very narrow molecular weight distribution: M_w/M_n ~ 1.04 to 1.06. Figure 2.4 shows the dependence between the melt index and the weight of the load for one such polymer with the molecular weight of 146,000 and the melt index I_2 of 0.48. When the weights are small, up to 5 kg, the relationship between the melt index and the load is linear; that is, the polymer melt behaves as a Newtonian liquid (Equation 2.17). This flow range provides the basis for the measurement of the polymer's zero-shear viscosity η_0 using Equation 2.17. When the weight of the load increases above 5 kg, an upward deviation from the linear dependence between the melt flow rate and the weight of the load becomes obvious, demonstrating the non-Newtonian nature of the melt at a high stress. However, the deviation from the Newtonian behavior for this monodisperse polymer is not large; the I_{21}/I_2 ratio for this polymer melt is merely 12.8, just slightly higher than the I_{21}/I_2 ratio for a Newtonian melt, which is equal to 10 by definition.

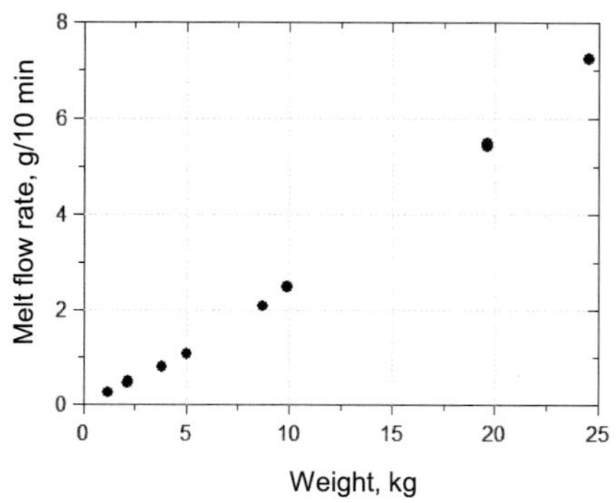

Figure 2.4
Melt flow data for monodisperse polyethylene with molecular weight of 146,000

This form of deviation from the Newtonian flow behavior has no universally accepted theoretical interpretation and it is usually expressed by *empirical* functions, such as [12, 17–19]:

$$\eta/\eta_0 = \left[1 + k_\eta \cdot (D \cdot \eta_0)^\beta\right]^{-1} \tag{2.18}$$

where D is the shear stress (see Equation 2.8) and k_η and β are two empirical parameters that are estimated from experimental data. Equation 2.18 signifies that as the shear stress increases, the intrinsic viscosity η of the melt decreases in comparison with its zero-shear viscosity η_0. The existence of such a decrease constitutes the definition of a non-Newtonian flow behavior. In everyday practice of melt index measurements, this phenomenon is usually called *shear thinning* of a polymer melt. The deviation from the Newtonian flow behavior for the monodisperse polymer melt in Figure 2.4 is too small to determine the k_η and β values with sufficient precision. More dependable estimations were carried out for complex polymer mixtures, as described below (Section 2.4.2). They give $k_\eta \approx 1 \cdot 10^{-5}$ and $\beta \approx 0.81$.

In principle, an analytical expression for the elementary flow rate dV_r at a radius r (similar to Equation 2.12) could be produced by introducing Equation 2.18 and Equation 2.8 into Equation 2.11. Continuing such a derivation would provide an analytical expression for the total flow rate V (m^3/sec) of a non-Newtonian melt consisting of macromolecules with the same polymerization degree. Unfortunately, such an analytical approach is not possible. Instead, a numerical solution of the problem was carried out. Some representative results are shown in Table 2.2. Melt indexes in the table were calculated from the computed flow rates V taking into account that the density of polyethylene melt at 190 °C is 0.764 g/cm^3; thus, I (g/10 min) = $4.58 \cdot 10^8 \cdot V$ (m^3/sec).

Table 2.2 Calculated Melt Indexes of Monodisperse Polyethylene Resins

Molecular weight:	25,000	50,000	75,000	100,000	200,000
η_0, kPa/sec:	0.54	5.29	21.9	59.9	665
I_2^{Newton}, g/10 min:	15.0	1.52	0.37	0.13	~0.010
I_2, g/10 min:	15.3	1.56	0.38	0.14	~0.012
I_{21}/I_2:	11.5	11.5	11.5	11.5	11.5

The results are clear: melts of monodisperse polymers, even when they have a quite high molecular weight, behave very similarly to Newtonian liquids. Their calculated I_{21}/I_2 ratios are about 11.5, that is, they are not significantly higher than the I_{21}/I_2 ratio for Newtonian melts (I_{21}/I_2 = 10), and their melt indexes I_2 are very close to the melt indexes of the respective Newtonian melts.

2.4 Additivity Rules for Melt Viscosity. Calculation of Melt Indexes and Melt Flow Ratios from Molecular Weight Distribution Data

Monodisperse polymers described in the previous section represent a rare exception in polymer chemistry. Most polymers, including all commercial polyethylene resins, are polydisperse, that is, they contain polymer molecules of many different lengths. The goal of the presentation in this section is to establish relationships between two sets of parameters: on one hand, the average molecular weight and the width of the molecular weight distribution of a polyethylene resin, and, on the other hand, the melt index and the melt flow ratio of the resin. Two subjects will be addressed: what is the zero-shear viscosity of a melt containing mixtures of macromolecules with different molecular weights and what is the effective viscosity of the same melt when it flows through a cylindrical capillary.

2.4.1 Additivity Rules for Zero-Shear Viscosity

The simplest example of a polydisperse polymer mixture is a blend of two monodisperse polymers, one with molecular weight MW_1 and the zero-shear viscosity $\eta_{0,1}$ and another with MW_2 and $\eta_{0,2}$, The polymers are mixed in the weight proportion *fraction* and 1 – *fraction*. What would the $\eta_{0,\text{mix}}$ value of such a mixture be? There is no good theoretical answer to this question, but several empirical mixing rules for the zero-shear viscosity have been proposed in the literature. One of the commonly used mixing rules is the $\log(\eta_0)$-additive rule [17–21]:

$$\log(\eta_{0,\text{mix}}) = \textit{fraction} \cdot \log(\eta_{0,1}) + (1 - \textit{fraction}) \cdot \log(\eta_{0,2}) \tag{2.19}$$

The second empirical mixing rule makes use of Equation 2.6. The underlying assumption is that the zero-shear viscosity of a binary polymer mixture can be estimated from the weight-average molecular weight of the mixture, M_w = *fraction* · MW_1 + (1 – *fraction*) · MW_2 [8, 22]. According to Equation 2.6, the polymerization degree of a monodisperse polymer is related to its zero-shear viscosity as $n = MW/28 = (\eta_{0,n}/k)^{-\gamma}$ where $\gamma = 3.5$. Therefore, if the M_w-mixing rule is applied to a mixture of two monodisperse polymers, the zero-shear viscosity of the mixture is:

$$\eta_{0,\text{mix}} = \left[\textit{fraction} \cdot (\eta_{0,1})^{-\gamma} + (1 - \textit{fraction}) \cdot (\eta_{0,2})^{-\gamma}\right]^{\gamma} \tag{2.20}$$

The applicability the two mixing rules in Equation 2.19 and Equation 2.20 for the estimation of the zero-shear viscosity of binary polymer mixtures was demonstrated experimentally. Figure 2.5 shows the experimental dependence between the melt index (the melt flow rate) and the load weight for the 1:1 mixture of two monodisperse polymers with the following parameters: MW_1 = 146,000, $\eta_{0,1}$ = 2850 Pa · sec and MW_2 = 51,000, $\eta_{0,2}$ ~ 450 · Pa · sec. The experimental zero-shear viscosity of the mixture was estimated from the melt flow rates at low loads: $\eta_{0,mix}$ ~ 1050 Pa · sec. The calculation with the log(η_0)-additive mixing rule gives $\eta_{0,mix}{}^{calc}$ = 1130 Pa · sec and the calculation with the M_w-mixing rule ~ 1250 Pa · sec. This comparison shows that both mixing rules are suitable for describing the η_0 data. The differences between the results of the two mixing rules are small and the use of a particular rule in the case of multicomponent polymer mixtures is determined purely by convenience of numerical calculations.

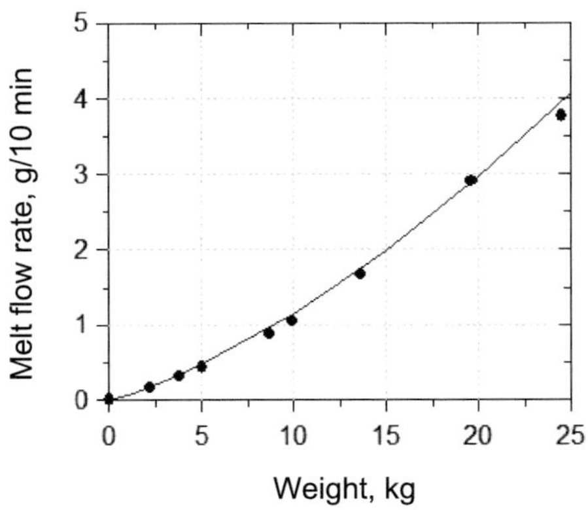

Figure 2.5
Melt flow of 1:1 mixture of two monodisperse polymers with MW_1 = 146,000 and MW_2 = 51,000

Of course, any real polyethylene resin contains not two monodisperse polymer components, as in the above example, but a very large number of macromolecules with different molecular weights MW_n and polymerization degrees n. The weight-average molecular weight for such a mixture is $M_w = \Sigma(fraction_n \cdot MW_n)$. For example, if the M_w-mixing rule is applied to the mixture, the calculation of its zero-shear viscosity $\eta_{0,mix}$ is based on the general form of Equation 2.20:

$$\eta_{0,\text{mix}} = \sum \left[fraction_n \cdot (\eta_{0,n})^{-\gamma} \right]^{\gamma} \tag{2.21}$$

with γ = 3.5.

The variable $fraction_n$ in Equation 2.21 is the weight fraction of polymer molecules with the polymerization degree n. Analytical expressions for the $fraction_n$ variable for polyethylene resins with a different molecular weight distribution are described in Section 1.4 and Section 1.5 of Chapter 1. Section 2.5 gives several examples of calculating the $fraction_n$ values.

2.4.2 Additivity Rules for Effective Viscosity and General Expressions for Flow of Multi-Component Non-Newtonian Melts

All commercially produced polyethylene resins are complex mixtures of macromolecules with a different polymerization degree, and their melts exhibit a pronounced non-Newtonian behavior [1–6, 23–30]. These melts can be viewed as assemblies of macromolecular coils with different radii. The rheological analysis of such polymer mixtures is a very challenging subject [28–30]. It is additionally complicated by the coil entanglement phenomenon: neighboring polymer coils penetrate each other and each coil is to a certain degree entangled with the coils next to it. The analysis of the melt behavior presented below is based on the simplified version of the rheological model which does not take into account the coil entanglement. This approach cannot provide a nuanced detailed picture of the polymer rheology. Instead, the simple model gives a semiquantitative description of the melt behavior, which assists in the physical interpretation of such basic (and easily experimentally measured) values as melt indexes and melt flow ratios.

The simple rheological model of a multicomponent melt is based on three conjectures:

1. In a given elemental volume of a flowing melt positioned at a distance r from the capillary axis (see Figure 2.1(b)), all the macromolecular coils are subjected to the same shear stress D_r.
2. The effective viscosity $\eta_{\text{mix},D}$ of a multicomponent melt in this elemental volume is governed by the same mixing rules as the mixing rules for the zero-shear viscosity $\eta_{0,\text{mix}}$ described in Section 2.4.1. For example, if the M_w-mixing rule is used, this conjecture is represented by the general equation similar to Equation 2.21:

$$\eta_{\text{mix},D} = \sum\left[fraction_n \cdot (\eta_{n,D})^{-\gamma}\right]^{\gamma} \tag{2.22}$$

   The variable $fraction_n$ in Equation 2.22 is the combined weight fraction of polymer molecules with the polymerization degree n, and $\eta_{n,D}$ is the effective viscosity of this polymer component at a shear stress D.

3. As the distance r of the elemental volume increases from 0 to R (Figure 2.1(b)), the shear stress D_r increases and the effective viscosity $\eta_{n,D}$ of each component in the polymer mixture is reduced due to the shear thinning effect according to Equation 2.18. The degree of deviation from the Newtonian behavior (the level of shear thinning) depends on the molecular weight of each polymer component (that is, on its zero-shear viscosity $\eta_{0,n}$): the higher the n value (and the higher $\eta_{0,n}$), the larger is the deviation.

This variability in the level of non-Newtonian properties depending on the polymerization degree explains a significant difference in the melt behavior of monodisperse polymers (Table 2.1) and multicomponent commercial resins. When the melt of a monodisperse polymer flows through a capillary, all of its macromolecules at a distance *r* from the capillary axis change their shape from spherical to ellipsoidal to the same degree (Figure 2.3); they all are equally elongated near the capillary wall and all retain the spherical shape in the center of the capillary where the shear stress is absent. In contrast, when the melt of a multicomponent polyethylene resin flows through a capillary, macromolecules with a higher molecular weight form more elongated ellipsoids whereas the shorter macromolecules retain a more spherical shape.

All commercial polyethylene resins are complex mixtures of polymer molecules with different molecular weights. Molecular weight distribution of any of these polymers can be represented as a combination of several Flory components with different weight-average molecular weights M_w. The number of Flory components in a particular resin depends on the type of catalyst used for the resin manufacture: one Flory component in polymers prepared with single-site catalysts, two Flory components in resins produced with supported metallocene catalysts, four to five Flory components in resins produced with supported Ziegler-Natta catalysts, and up to eight Flory components in resins produced with bicomponent and chromium oxide catalysts.

In the general case, a polymer with a multi-component molecular weight distribution is characterized by three parameters:

1. The number of Flory components, j.
2. The weight fraction of each Flory component, $\boldsymbol{FR}_j$ ($\Sigma \boldsymbol{FR}_j = 1$).
3. The average molecular weight of each Flory component, $\boldsymbol{M}_{w,j}$, and its number-average polymerization degree $\mathbf{n}_{av,j}$ ($\mathbf{n}_{av,j} = \boldsymbol{M}_{w,j}/56$).

The combined weight fraction of all macromolecules containing *n* ethylene units (the $fraction_n$ variable in Equation 2.22) in such a polymer is:

$$fraction_n = \sum \boldsymbol{FR}_j \cdot \delta(E)_{n,j} = \sum \boldsymbol{FR}_j \cdot \left[(n / \mathbf{n}_{av,j}^2) \cdot \mathrm{Exp}(-n / \mathbf{n}_{av,j}) \right] \tag{2.23}$$

If the M_w-additivity rule for the viscosity of a multi-component melt is used (Equation 2.20), then the zero-shear viscosity of this polymer mixture is (from Equation 2.21 and Equation 2.6):

$$\eta_0^{\text{multi-Flory}} = \sum \boldsymbol{FR}_j \cdot \int \left[\left(n / \mathbf{n}_{av,j}\right)^2 \cdot \mathrm{Exp}\left(-n / \mathbf{n}_{av,j}\right) \cdot k^{-\gamma} \right]^{\gamma} \cdot dn \tag{2.24}$$

with $\gamma = 3.5$.

Finally, the expression for the effective viscosity $\eta_D^{\text{multi-Flory}}$ of the multi-Flory polymer mixture subject to shear stress D_r is produced by combining three equations, the expression for the zero-shear viscosity of a multi-Flory polyethylene resin in Equation 2.24, the M_w-additivity rule for the effective viscosity in Equation 2.22, and the shear-thinning rule in Equation 2.18:

$$\eta_D^{\text{multi-Flory}} = \sum_i \boldsymbol{FR}_j \cdot \int \Big[(n/\mathbf{n}_{\text{av,j}})^2 \cdot \text{Exp}(-n/\mathbf{n}_{\text{av,j}}) \cdot \left\{ k \cdot n^{\gamma} \cdot \left[1 + k_{\eta} \cdot (D_r \cdot k \cdot n^{\gamma})^{\beta}\right]^{-1} \right\}^{-\gamma} \Big]^{\gamma} \cdot dn \qquad (2.25)$$

Equation 2.25 is too cumbersome to be solved analytically in the form "$\eta_D^{\text{multi-Flory}}$ as a function of n" but it can be solved numerically. To carry out the numerical integration, the integration limits in Equation 2.21 and Equation 2.25, from 0 to ∞, are replaced with practical adequate limits, for example, from $0.05 \cdot \mathbf{n}_{\text{av,j}}$ to $10 \cdot \mathbf{n}_{\text{av,j}}$. Table 2.3 lists computational steps required to calculate melt velocity profiles and melt flow rates (melt indexes) of resins with a multi-Flory molecular weight distribution.

Table 2.3 Computational Steps in the Procedure for Calculating Melt Flow Velocity Profile and Melt Flow Rate (Melt Index) of Polyethylene Resin with Multi-Flory Molecular Weight Distribution

First step	Molecular weight distribution of a resin is represented as a combination of several (**j**) Flory components with fractions $\boldsymbol{FR}_j$ and weight-average molecular weights $\boldsymbol{M}_{w,j}$ – see Section 1.5 in Chapter 1.
Second step	The zero-shear viscosity of this multi-Flory polymer, $\eta_0^{\text{multi-Flory}}$, is calculated with Equation 2.24.
Third step	The effective viscosity of the multi-Flory polymer under stress D_r, $\eta_D^{\text{multi-Flory}}$, is calculated numerically with Equation 2.25 for a number of capillary radii r from 0 to R in the form "$\eta_D^{\text{multi-Flory}}$ as a function of D_r".
Fourth step	The principal equation for the melt flow of a multi-Flory polymer through a capillary, $(P/2L) \cdot r = \eta_D^{\text{multi-Flory}} D_r$ (analog of Equation 2.11), is solved numerically in the form "D_r as a function of r".
Fifth step	Numerical results of the previous step are approximated in a mathematical form convenient for integration.
Sixth step	The melt flow velocity v_r is calculated as a function of r using the definition of D_r, $D_r = -dv_r/dr$ (Equation 2.8) under the boundary condition $v_R = 0$.
Seventh step	The melt flow rate of the multi-Flory polymer through a capillary, V (m^3/sec), is calculated with Equation 2.13. Finally, taking into account that the melt density of polyethylene at 190 °C is 0.764 g/cm^3, the melt index of the multi-Flory polymer is calculated as I (g/10 min) = $4.58 \cdot 10^8 \cdot V$ (m^3/sec).

As an example, a convenient form for approximating the results of the numerical integration in the fourth step in Table 2.3 is "$D_r = \mathbf{K}_1 \cdot r + \mathbf{K}_2 \cdot r^2 + \mathbf{K}_3 \cdot r^3$". In this case, after the values of $\mathbf{K}_1$, $\mathbf{K}_2$ and $\mathbf{K}_3$ are estimated, the linear melt flow velocity is calculated as v_r (m/sec) $= \mathbf{K}_1 \cdot (R^2 - r^2)/2 + \mathbf{K}_2 \cdot (R^3 - r^3)/3 + \mathbf{K}_3 \cdot (R^4 - r^4)/4$, and the volume melt flow rate is calculated as V (m^3/sec) $= \pi(\mathbf{K}_1 \cdot R^4/4 + \mathbf{K}_2 \cdot R^5/5 + \mathbf{K}_3 \cdot R^6/6)$.

The simplest example of this calculation sequence is presented in Figure 2.5. The experimental data (points) in the plot show the rheological behavior of a 1:1 mixture of two monodisperse polymers with different molecular weights. The calculation starts with the second step in Table 2.3, the estimation of the zero-shear viscosity for the binary mixture, as described in Section 2.4.1. After that, simplified versions of steps 3 to 7 are performed. The final computation result, the solid line in Figure 2.5, represents the experimental data quite well: the binary polymer mixture exhibits a more expressed non-Newtonian rheological behavior than each of its monodisperse components (compare Figure 2.4 and Figure 2.5).

2.5 Examples of Melt Flow Behavior of Commercial Polyethylene Resins

The principal difficulty in using the simple theory of the melt flow outlined in Section 2.4 is that its three parameters, k in Equation 2.6 and k_η and β in Equation 2.18, are empirical and must be determined from experiment. Molecular coils in polymer melts are the same in size in the melts of monodisperse polymers but their radius of gyration differs greatly in melts of commercial resins with a multi-Flory molecular weight distribution (Equation 2.5). As a result, the values of parameters k, k_η, and β may differ depending on the degree of interaction between neighboring coils of different radii and on proportions between long, medium-length, and short macromolecules in a polymer mixture. In addition, the radius of gyration of a polyethylene coil depends on the degree of chain branching (the content of α-olefin in a copolymer molecule), which is often different in copolymer molecules of different molecular weight.

Rheological calculations in the following sections use the estimation of the k parameter from the experimental melt flow data for metallocene polyethylenes. It ranges from $7.0 \cdot 10^{-9}$ to $8.5 \cdot 10^{-9}$ kPa · sec. This k value is suitable for melt flow characterization of typical HDPE and MDPE resins in a broad I_2 range, from ~ 1 to >10 g/10 min. In the case of LLDPE resins, the k value increases to $1.3 - 1.5 \cdot 10^{-8}$ kPa · sec. Parameters k_η and β were also estimated from the melt flow data for metallocene polyethylenes; at 190 °C they are: $k_\eta \approx 5.0 \cdot 10^{-6}$, $\beta \approx 0.85$, which is close to the estimation for the mixture of two monodisperse ethylene polymers in Figure 2.5 ($k_\eta = 6.5 \cdot 10^{-6}$, $\beta = 0.87$).

When a new type of a resin is produced, it is always much more reliable and much faster to measure the I_2 and I_{21} values experimentally rather than analyze the molecular weight distribution of the resin by the GPC method and then calculate its I_2 and I_{21} values. The principal application of the rheological model in Section 2.4 is the interpretation of experimentally measured melt flow characteristics of various resins and the prediction of possible changes in the melt properties of resins produced with modified catalysts.

2.5.1 Polyethylene Resins Produced with Single-Site Metallocene Catalysts

Polyethylene resins with a single-Flory molecular weight distribution are relatively rare; they can be prepared with metallocene catalysts at very high [MAO]/[metallocene complex] ratios under strictly controlled laboratory conditions, as described in Section 1.5 of Chapter 1. These polymers represent the simplest example of the rheological behavior of multicomponent resins produced in catalytic polymerization reactions. The molecular weight distribution of these polymers is described by one Flory function with a single parameter, the number-average polymerization degree $\mathbf{n}_{av} = M_w/56$; the M_w/M_n ratio for these polymers is equal to 2. When rheological properties of single-Flory polymers are calculated, Equation 2.22 to Equation 2.25 are simplified: the term $\boldsymbol{FR}_j$ in them is set to 1 and the expression for the weight fraction of macromolecules containing n ethylene units (Equation 2.22) is reduced to $fraction_n = (n/\mathbf{n}_{av}^2) \cdot \mathrm{Exp}(-n/\mathbf{n}_{av})$. Even after these replacements, the expression for $\eta_D^{\text{single-Flory}}$ (the simplified form of Equation 2.25) cannot be solved analytically and the computational steps detailed in Table 2.3 should be followed to calculate melt velocity profiles and melt indexes of the resins.

Table 2.4 gives the results of melt index calculations for five single-Flory polyethylene resins in a broad range of average molecular weights. The results are very informative. When the weight of the load is small, 2.16 kg, the I_2 values of these non-Newtonian melts are only slightly higher than the I_2^{Newton} values calculated under the assumption that the melts exhibit the Newtonian behavior; the differences do not exceed 7 to 8%. However, the I_{21} values of the melts are significantly higher than the I_{21}^{Newton} values (by the definition of a Newtonian melt, $I_{21}^{\text{Newton}} = 10 \cdot I_2^{\text{Newton}}$). The differences between I_{21} and I_{21}^{Newton} are of the order of 40%. The calculated melt flow ratios for the single-Flory polymers, $I_{21}/I_2 \sim 16$, match the experimentally measured I_{21}/I_2 ratios for polyethylene resins produced with single-site metallocene catalysts, 15 to 16.

Table 2.4 Calculated Melt Indexes of Polyethylene Resins with Single-Flory Molecular Weight Distribution

M_w:	50,000	75,000	100,000	125,000	150,000
η_0^{Flory}, kPa/sec:	0.46	1.92	5.25	11.5	21.7
I_2^{Newton}, g/10 min:	17.3	4.18	1.53	0.70	0.37
I_2, g/10 min:	18.7	4.51	1.65	0.76	0.40
I_{21}^{Newton}, g/10 min:	173	41.8	15.3	7.0	3.7
I_{21}, g/10 min:	298	71.9	26.3	12.1	6.4
I_{21}/I_2:	15.9	15.9	15.9	15.9	15.9

Figure 2.6 compares flow velocity profiles of two melts, Newtonian and non-Newtonian, calculated with Equation 2.8 for a polyethylene resin with the single-Flory molecular weight distribution under the conditions of I_2 and I_{21} measurement. The flow velocity v_r of both melts is equal to zero at the capillary wall (at capillary radius r = 1.027 mm) and it reaches a maximum at the capillary axis, at r = 0. When the weight of the load is small, (Figure 2.6(a)), flow velocity profiles for the Newtonian melt and the non-Newtonian melt are very close, but the non-Newtonian melt flows much faster compared to the Newtonian melt when the weight of the load is high (Figure 2.6(b)), demonstrating the shear-thinning phenomenon in multi-component melts.

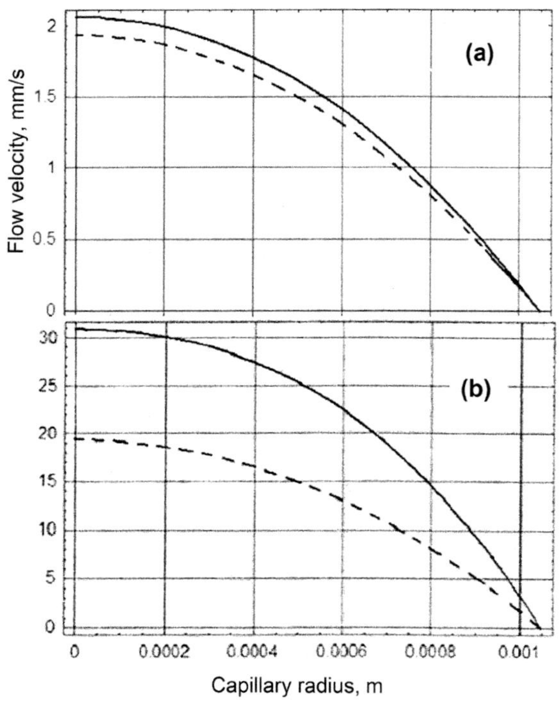

Figure 2.6
Melt flow profiles of single-Flory polymer with M_w = 100,000 at two loads, 2.16 kg (a) and 21.6 kg (b); solid lines are for non-Newtonian melt and dashed lines are for Newtonian melt

2.5.2 LLDPE Resins Produced with Supported Metallocene Catalysts

Most commercial polyethylene resins manufactured with metallocene catalysts have a broader molecular weight distribution than that described by the single Flory function (Equation 1.9). Their GPC curves can be adequately represented as a sum of GPC curves of two Flory components (Figure 1.5) with average molecular weights differing by a factor of ~2. Resins with such a molecular weight distribution are produced, for example, with metallocene catalysts supported on porous inorganic carriers (see an example in Figure 1.4). These catalysts have two types of active center which coexist on the catalyst internal surface and each produces a fraction of the bi-Flory polymer blend.

Table 2.5 shows rheological properties of two commercial LLDPE resins with a bi-Flory molecular weight distribution produced with a supported metallocene catalyst, the resin formulated for the manufacture of cast film and the resin formulated for the manufacture of blown film. The table lists two sets of data for each resin. The first set includes parameters of the polymer blends determined from the GPC data: the average molecular weight of each Flory component, $\boldsymbol{M}_{w,1}$ and $\boldsymbol{M}_{w,2}$, the content of the low molecular weight Flory component in the blend, $\boldsymbol{FR}_1$, and average molecular weight parameters of the blends, M_a^{ws} and M_w/M_n ratios. The second set of the data gives calculated rheological properties of both resins and their Flory components: the calculated zero-share viscosity of each Flory component and the blend, η_0 values, the melt index I_2 and the melt flow ratio I_{21}/I_2.

Table 2.5 Rheological Properties of Two Polyethylene Resins with Bi-Flory Molecular Weight Distribution

Cast-film resin, I_2 = 3.5 g/10 min								
Component	Experimental data					Calculation		
	$M_{w,1}$	$M_{w,2}$	FR_1	M_w^{av}	M_w/M_n	η_0, kPa·sec	I_2, g/10 min	I_{21}/I_2
Total resin	41,900	88,300	0.54	65,300	2.29	2.54	3.6	16.1
Low-MW component	41,900		-	-	(2.0)	0.77	11.5	15.0
High-MW component		88,300	-	-	(2.0)	10.3	0.86	15.1
Blown-film resin, I_2 = 1.2 g/10 min								
Total resin	53,000	117,000	0.46	87,600	2.33	7.73	1.20	16.2
Low-MW component	53,000		-	-	(2.0)	1.74	5.09	15.0
High-MW component		117,000	-	-	(2.0)	27.5	0.32	15.0

The molecular weight distribution data show that both resins are about 1:1 mixtures of two closely positioned Flory components (Figure 1.4). The M_w/M_n ratios of the combined products, ~ 2.3, are only slightly higher than the M_w/M_n ratio of polymers with a single-Flory molecular weight distribution. However, rheological properties of the Flory components in each resin are very different. If these polymer components were synthesized separately, zero-shear viscosities and melt indexes of the two materials would differ by a factor ~ 13 to 16.

2.5.3 LLDPE Resins Produced with Supported Ziegler-Natta Catalysts

LLDPE resins produced with supported Ziegler-Natta catalysts have a medium-broad molecular weight distribution; their I_{21}/I_2 ratios range from 27 to 32 depending on the catalyst recipe and polymerization conditions. GPC curves of all these resins contain four to five Flory components.

This section describes rheological features of one such resin, an ethylene/1-hexene copolymer containing 3.3 mol % of 1-hexene. Technological parameters of the polymer are typical for LLDPE resins of this type: density = 0.917 g/cm^3, M_w ~ 102,000, M_w/M_n = 4.9, I_2 = 1.1 g/10 min, and I_{21}/I_2 = 28.5. Table 2.6 provides information on the molecular weight distribution of the resin derived from its GPC analysis. The polymer consists of four Flory components with average molecular weights varying in a broad range, from ~ 14,000 to over 300,000.

Table 2.6 Flory Components in LLDPE Resin Produced with Supported Ziegler-Natta Catalyst

Flory component:	I	II	III	IV
Molecular weight, $\boldsymbol{M_{w,j}}$:	14,300	43,400	110,700	303,400
Fraction, $\boldsymbol{FR_j}$:	0.12	0.37	0.35	0.15

Figure 2.7 shows experimental rheological characteristics of the resin measured with an extrusion plastometer, the dependence between the weight of the load and the melt flow rate through the standard capillary (the melt index). These data demonstrate a strongly expressed non-Newtonian behavior of the melt; the dependence between the melt flow rate and the load weight should be linear for Newtonian melt (Equation 2.17).

The computational procedure described in Table 2.3 was used to model the rheological behavior of the resin using the molecular weight data for the Flory components from Table 2.6. The modeling results are also shown in Figure 2.7. The comparison of the experimental data (points) and the calculation (line) shows that the rheological model detailed in Section 2.4.2 is quite adequate for the description of the non-Newtonian rheological behavior of multicomponent polymer mixtures. The calculations also provide detailed information about the contributions of different Flory components to the rheological behavior of the resin. The results are shown in Table 2.7.

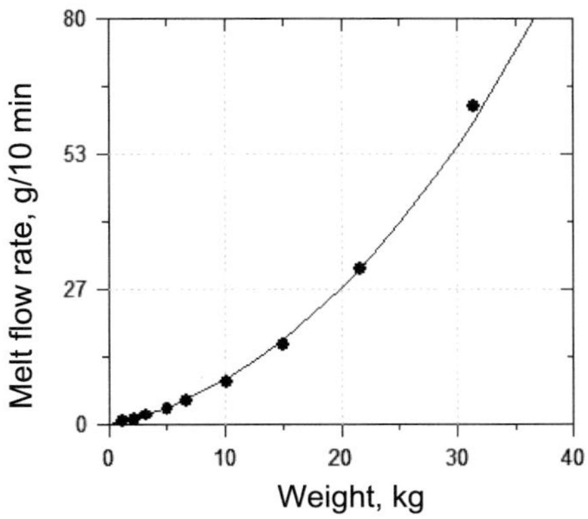

Figure 2.7
Melt flow rate of a multi-Flory LLDPE resin prepared with Ziegler-Natta catalyst as a function of load weight; experimental data (●) and theoretical simulation (line)

Table 2.7 Rheological Parameters of Flory Components in LLDPE Resin Produced with Supported Ziegler-Natta Catalyst

Flory component:	I	II	III	IV	Total resin
Molecular weight, $M_{w,j}$:	14,300	43,400	110,700	303,400	102,000
Calculated rheological parameters of individual Flory components and the total resin					
η_0, kPa·sec:	0.011	0.49	12.9	430	13.0
I_2, g/10 min:	~760	18	0.7	0.02	1.0

The computational results show that this resin is a mixture of polymer components with a very different zero-shear viscosity, from ~ 0.01 kPa · sec for component I to over 400 kPa · sec for component IV. If the Flory component IV were synthesized separately, such a resin could not be processed with modern equipment. On the other hand, the Flory component I is a material of very low viscosity; if produced separately, it would have the melt index of over 750 g/10 min. Fractionation data in Table 1.3 also show that this component has very a high hexene content; it is a completely amorphous sticky material of little commercial value. This material does not contribute anything to the mechanical strength of the polymer mixture; its only purpose is to dilute components III and IV and make the polymer mixture processable.

The data in Table 2.7 also demonstrate several aspects of the shear thinning phenomena in multicomponent polyethylene melts. The calculated zero-shear viscosity η_0, of this resin is ~ 13 kPa/sec. If this melt had Newtonian rheological properties, its I_2 melt index would be 0.62, which is ~40% lower compared to the real I_2 value (1.1) and the calculated I_2 value, ~1. Even when the melt pressure is low (the load weight 2.16 kg), this melt flows noticeably faster than the melt of a Newtonian liquid with the same zero-shear viscosity would. The difference becomes even more pronounced when the I_{21} values are compared: the calculated I_{21} value of the resin is 30.9 g/10 min whereas the I_{21} value of the respective Newtonian melt would be 6.2 g/10 min. Still another measure of shear thinning of this melt comes from the comparison of maximum velocities of the melt in the center of the capillary. The ten-fold increase in the load weight, from 2.16 to 21.6 kg, produces a nearly 27-fold increase of the maximum velocity, from 1.3 to ~34.5 mm/sec. This velocity ratio correlates with melt flow ratio of the resin, $I_{21}/I_2 = 28.5$.

The model outlined in Section 2.4.2 is also useful in the analysis of changes one can expect in the rheological behavior of polyethylene resins when conditions of their synthesis are changed or when the catalyst recipe is modified. Several examples of these changes are presented in Table 2.8. The data in columns 2 and 3 describe the effect of reaction conditions, a change in the concentration of hydrogen in a polymerization reactor by ±10%. This change leads a proportional change in the molecular weight of each Flory component. The calculations show that such changes do not affect the width of the molecular weight distribution of the resin; the M_w/M_n ratio and the melt flow ratio of the resin remain unchanged. However, the changes in the hydrogen concentration greatly affect both the I_2 and the I_{21} values. These data emphasize importance of a precise control of reaction parameters in commercial processes: a 5 to 7% drift in some process parameters could lead to the production of an off-spec. material.

Commercial manufacturers of polyethylene resins periodically attempt to modify recipes of supported Ziegler-Natta catalysts with the goal of producing resins of better quality and a higher commercial value. The last two columns of Table 2.8 show expected rheological properties of polymers produced with two hypothetical modified catalysts. As mentioned above, Flory component I has very a very low molecular weight (Table 2.6) and a high α-olefin content (Table 1.3). Some types of chemical modification allow a nearly complete elimination of the active centers in the catalysts which produce this material. Unfortunately, such a modification is usually accompanied by a significant reduction in the yield of Flory component II as well. The example in the fifth column of Table 2.8 describes rheological consequences of this catalyst modification; it shows rheological properties of a resin produced under the same reactor condition as the original resin but when the whole component I and one-half of component II are absent from the polymer. The results of the modification are as expected: the new polymer mixture with missing low molecular weight components has a narrower molecular weight distribution (the M_w/M_n ratio is reduced from 4.9 to 3.8), a higher zero-shear viscosity, a lower I_2 value, and a lower melt flow ratio, 26.4 instead of 28.5 for the resin produced with the original catalyst.

Table 2.8 Calculated Melt Indexes of and Melt Flow Ratios of Several LLDPE Resins Produced with Ziegler-Natta Catalyst

Resin:	Original resin	Molecular weight increased	Molecular weight decreased	Flory components I & II decreased	Flory component IV decreased
M_w:	102,000	113,000	92,700	109,000	85,800
M_w/M_n:	4.9	4.9	4.9	3.8	3.8
η_0, kPa/sec:	13	~300	9.2	16.4	7.1
I_2, g/10 min:	1.0	0.75	1.51	0.81	1.90
I_{21}/I_2:	28.5	28.5	28.5	26.4	26.2

The example in the last column of Table 2.8 shows the consequences of another attempt at catalyst modification. Flory component IV has very low 1-hexene content, below 0.4 mol % (Table 1.3). This is a highly crystalline material with a very high molecular weight (Table 2.6). If the fraction of this component in the polymer mixture is reduced by one-half, the remaining material is expected to have a higher average 1-hexene content, a lower average molecular weight, and a narrower molecular weight distribution. Table 2.8 shows that the rheological consequences of this catalyst modification are significant as well: the melt index of the resin nearly doubles and its I_{21}/I_2 value decreases by two units. Although both types of catalyst modification produce the same overall effect, formation of LLDPE resins with a lower I_{21}/I_2 ratio, two different corrective measures should be carried out in the respective polymerization conditions. If centers I and II are poisoned, the copolymerization reaction should be carried out at a higher α-olefin concentration in the reactor to compensate for the loss of these Flory components and to bring the average α-olefin content in the copolymer mixture to the original level, ~3 mol %. If the amount of component IV is reduced, the 1-hexene concentration in the reactor should be reduced to decrease the average α-olefin content in the resin and to avoid the formation of excessive amounts of low-viscosity components I and II.

2.5.4 HDPE Resins with Bimodal Molecular Weight Distribution

All polyethylene resins with a bimodal molecular weight distribution are mixtures of two fractions, one with a very high molecular weight and another with a very low molecular weight. Molecular weight distribution curves of these resins consist of two clearly identifiable components. Two types of bimodal resins are produced commercially. Resins of the first type are synthesized with bimetallic catalysts containing active centers of two different types. One of the catalyst components is a metallocene or a post-metallocene catalyst, which produces the low molecular weight fraction of the bimodal resin, and another catalyst component is a Ziegler-Natta catalyst, which produces its high molecular weight fraction. The second type of bimodal resins is produced with a single Ziegler-Natta catalyst in two reactors connected in a series and operating under different reaction conditions (see Section 1.2).

Figure 1.8 in Chapter 1 shows a typical GPC curve of a bimodal HDPE resin produced with a bicomponent Ziegler-Natta/metallocene catalyst. Table 2.9 lists the data on molecular weights for both polymer fractions. Each fraction by itself has a relatively narrow molecular weight distribution. The metallocene-derived fraction has three closely spaced Flory components (M_w/M_n = 2.9), and the Ziegler-Natta-derived fraction has four Flory components (M_w/M_n = 6.0). However, molecular weights of the two polymer fractions differ greatly: the M_w value for the low molecular weight fraction is ~10,000 and that for the high molecular weight fraction is ~640,000. As a result, the M_w/M_n ratio for the total resin is very high, ~50.

Table 2.9 Bimodal Pipe-Grade HDPE Resin Produced with Bicomponent Ziegler-Natta/Metallocene Catalyst; Details of Molecular Weight Distribution

	M_w	M_w/M_n	Content, %
Total polymer:	352,200	~50	100
Low molecular weight fraction produced with metallocene catalyst component			
Total fraction:	9,800	2.9	46.0
Flory component I:	3,430	(2.0)	9.1
Flory component II:	6,170	(2.0)	18.4
Flory component III:	16,570	(2.0)	18.5
High molecular weight fraction produced with Ziegler-Natta catalyst component			
Total fraction:	644,400	6.0	54.0
Flory component I:	76,360	(2.0)	13.0
Flory component II:	230,500	(2.0)	15.3
Flory component III:	588,600	(2.0)	12.4
Flory component IV:	1,729,500	(2.0)	13.3

Table 2.10 gives calculated rheological properties of the total pipe-grade resin and its fractions. In rheological terms, the two fractions are very different. Their zero-shear viscosity differs by a factor of $3.4 \cdot 10^6$ and their high-load melt indexes I_{21} by a factor of $4.7 \cdot 10^7$. The high molecular weight fraction is very viscous; if produced separately, its I_{21} value of ~0.08 would be barely measurable. Such resins cannot be processed with modern extruders. On the other hand, the low molecular weight fraction is practically a wax; its viscosity at 190 °C would not differ significantly from the viscosity of water. This resin fraction does not contribute anything to the mechanical strength and toughness of the bimodal resin; its only purpose is

to serve as a diluent for the high molecular weight fraction during those several minutes when the resin melt is processed into a pipe.

Table 2.10 Rheological Parameters of Pipe-Grade HDPE Resin Produced with Bicomponent Ziegler-Natta/Metallocene Catalyst

Rheological parameters:	η_0, kPa/sec	I_2, g/10 min	I_{21}, g/10 min	I_{21}/I_2
Experimental values				
Total polymer:	–	0.045	6.2	139
Calculated values				
Total polymer:	~1000	0.052	7.2	140
Low molecular weight fraction (33%):	$2.4 \cdot 10^{-3}$	$4.9 \cdot 10^{3}$	$1.5 \cdot 10^{5}$	~30
High molecular weight fraction (67%):	$8.3 \cdot 10^{3}$	$1.9 \cdot 10^{-3}$	~0.08	~40

Figure 2.8 compares melt flow profiles of two resins through the capillary of an extrusion plastometer at the 2.16 kg load weight. The first resin is the bicomponent resin with the rheological parameters listed in Table 2.10 and the second resin is a Newtonian liquid with the same zero-shear viscosity, η_0 ~ 1000 kPa/sec. This comparison clearly demonstrates that the rheological behavior of resins with a broad bimodal molecular weight distribution differs greatly from the melt behavior of resins with Newtonian rheological properties, even under relatively small loads.

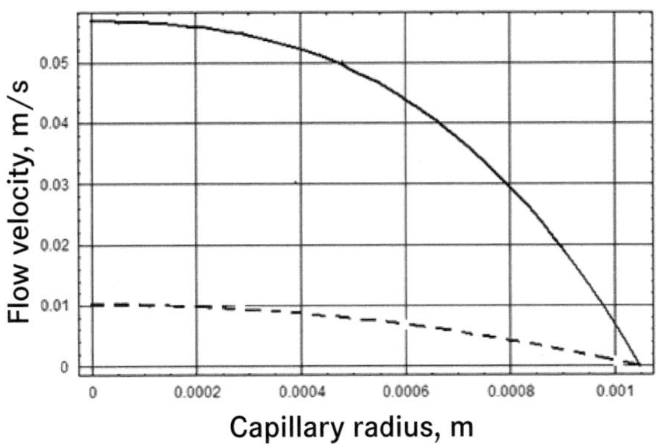

Figure 2.8
Melt flow profiles of resins with bimodal molecular weight distribution at 2.16 kg load; solid line is for non-Newtonian melt and dashed line is for Newtonian melt

Both Figure 2.8 and Figure 2.6(b) show another characteristic feature of polymer melts with a strongly expressed non-Newtonian behavior: a large velocity gradient near the wall of the capillary, at $r \approx 1$ mm. All rheological models of viscous fluids, both Newtonian and non-Newtonian, assume that the flow velocity at the wall of the capillary is zero (Figure 2.1(b)); this number is used as the low integration limit in Equation 2.14. This assumption usually holds in a broad range of loads.

However, when the loads are very high, this rule sometimes fails and a melt starts slipping along the capillary wall. The slipping usually proceeds in a stop-and-go manner, and as a result, the surface of the outflowing melt, after it solidifies, exhibits a specific rough pattern usually called "shark skin". The development of the shark-skin surface on an extrudate makes all melt index measurements (and all melt index calculations) meaningless. When the melt starts slipping along the capillary wall, the amount of the extrudate becomes dependent not only on the load weight but on the smoothness of the capillary surface as well; the smoother the surface, the higher is the load weight leading to the melt slipping and the development of the shark skin effect.

The second type of bimodal polyethylene resin is produced with the use of a single Ziegler-Natta catalyst in two slurry or gas-phase reactors connected in a series and operating under different conditions; at different temperatures and at different concentrations of hydrogen. Figure 1.9 gives an example of the molecular weight distribution of such a bimodal HDPE resin. Polymers of this type are also blends of two fractions, each consisting of four Flory components and shifted by a factor of ~20 in their average molecular weight (see Table 1.6). Table 2.11 gives calculated rheological parameters of this resin. Overall, the results of the rheological analysis for the two types of bimodal resins in Table 2.10 and Table 2.11 are similar. In the latter example (Table 2.11), the zero-shear viscosity of the two fractions differs by a factor of $3.8 \cdot 10^5$ and their high-load melt index I_{21} by a factor of $2.4 \cdot 10^6$. To take advantage of excellent mechanical properties of the high molecular weight material in both polymer mixtures and allow the resin processing, it must be diluted with a polymer of a very low molecular weight and very low viscosity.

Table 2.11 Rheological Parameters of Bimodal HDPE Resin Produced with Ziegler-Natta Catalyst in Two-Reactor Process

Rheological parameters:	η_0, kPa/sec	I_2, g/10 min	I_{21}, g/10 min	I_{21}/I_2
Total polymer:	130	0.20	15.6	~77
Low molecular weight fraction (45%):	0.024	558	1.8·105	~32
High molecular weight fraction (55%):	~910	0.014	0.40	~29

2.5.5 Effect of Long-Chain Branching

Three types of polyethylene resins have long-chain branches in their macromolecules: LDPE resins produced via the radical mechanism in high-pressure polymerization reactors, some MDPE and HDPE resins prepared with special types of metallocene catalysts, and HDPE resins produced with chromium oxide catalysts. To be rheologically significant, such branches must be longer than the critical entanglement length, which is equal to ~140 carbon atoms in the case of polyethylene [31, 32].

Effects of long-chain branching on the rheology of polyethylene resins have been studied both experimentally and theoretically [11, 27, 29] but still remain a very complicated subject. The single empirical rule for the description of such an effect is: the higher the content of long branches in macromolecules, the stronger is the deviation of the melts from the standard rheological behavior. In particular, the higher the content of long branches, the higher are the melt flow ratios of such resins compared to the melt flow ratios of resins of the same molecular weight but lacking long-chain branches.

Calculations of the I_{21}/I_2 ratios for a typical HDPE resin prepared with a chromium oxide catalyst can serve as an example of the long-chain branching effect. All these resins have a broad molecular weight distribution and their GPC curves have a characteristic triangular shape shown in Figure 1.7. Parameters of Flory components in one such resin are listed in Table 1.4. The average molecular weight of these resins is usually 220,000 to 250,000 and the M_w/M_n ratio is in the 12 to 15 range. The calculated zero-shear viscosity of the resin with the GPC curve in Figure 1.7 is 213 kPa/sec. A Newtonian melt with this viscosity would have the I_{21} value of 0.38 g/10 min but the resin has a broad molecular weight distribution and should exhibit an expressed non-Newtonian behavior. Rheological calculations described in Section 2.4 give the following estimations: I_{21} would be ~ 5.6 g/10 min (which is 15 times higher than the I_{21} value for a Newtonian melt with the same zero-shear viscosity) and the calculated I_{21}/I_2 ratio would be ~45. However, experimentally measured I_{21}/I_2 ratios for HDPE resins produced with chromium oxide catalysts are usually much higher; they range between 120 and 150. These large differences between the experimental and the calculated I_{21}/I_2 values are a clear manifestation of the long-chain branching effect on the rheological behavior of polyethylene resins.

References

1. Bremner, T., Rudin, A., Cook, D. G., *J. Appl. Polym. Sci.* (1990) 41, p. 1617
2. Bremner, T., Cook, D. G., Rudin, A., *J. Appl. Polym. Sci.* (1991) 43, p. 1773
3. Seavey, K. C., Liu, Y. A., Khare, N. P., Bremner, T., Chen, C.-C., *Ind. Eng. Chem. Res.* (2003) 42, p. 5354
4. Azmi, A., Sata, S. A., Rohman, F. S., Aziz, N., *J. Phys.: Conf. Ser.* (2019) 1349, p. 12094
5. Teh, J., Lam, P., Dobbin. C., *Polym. Test.* (2015) 47, p. 101
6. Balke, S. T., Mourey, T. H., Lusignan, C. P., *Int. J. Polym. Anal. Charact.* (2006) 11, p. 755
7. Kissin, Y. V., Nowlin, T. E., Wagner, K. P., *J. Polym. Sci., Part A: Polym. Chem.* (1988) 26, p. 755
8. Kissin, Y. V., Brandolini, A. J., Mink, R. I., Nowlin, T. E., *J. Polym. Sci., Part A: Polym. Chem.* (2009) 47, p. 3271

9. Lai, S. Y., Wilson, J. R., Knight, G. W., Stevens, J. C., Chum, P. S., *U. S. Patent* 5 272 236 (1993)
10. Knight, G. W., Lai, S., *SPE Polyolefins VIII International Conference* (1993) Houston, TX, p. 226
11. Seung, J. P., Larson, R. G., *J. Rheol.* (2005) 49, p. 523
12. Larson, R. G., *The Structure and Rheology of Complex Fluids* (2001) Oxford University Press, New York
13. Bird, R. B., Armstrong, R. C., Hassager, O., *Dynamics of Polymer Fluids*, vol. 1 (1987) John Wiley, New York, p. 105
14. Dealy, J. M., Read, D. J., Larson, R. G., *Structure and Rheology of Molten Polymers: From Structure to Flow Behavior* 2nd ed. (2018) Hanser, Munich
15. Oka, S., In *Rheology, Theory and Applications*, vol. 3. Eirich, F. R. (Ed.) (1960) Academic Press, New York, p. 21
16. Coolbaugh, T. S., Loveless, F. C., Matthews, D. N., *U. S. Patent* 5 276 100 (1994)
17. Larson, R. G., Qiang, Z., Sachin, S., Seung, J. P., *AIChE J.* (2007) 53, p. 542
18. Vinogradov, G. V., Malkin, A. Y., *J. Polym. Sci., Part A-2* (1966) 4, p. 135
19. Bondi, A., In *Rheology, Theory and Applications*, vol. 4. Eirich, F. R. (Ed.) (1967) Academic Press, New York, p. 70
20. Utracki, L. A., *Polymer Alloys and Blends. Thermodynamics and Rheology* (1990) Hanser, Munich, Part 3.3.1
21. Ghasemi, S.M., Sadeghi, G. M. M., *J. Appl. Polym. Sci.* (2008) 108, p. 2988
22. Muñoz-Escalona, A., Lafuente, P., Vega, J. F., Muñoz, M. E., Santamaria, A., *Polymer* (1997) 38, p. 589
23. Dealy, J. M., Larson, R. G., *Structure and Rheology of Molten Polymers: From Structure to Flow Behavior* (2006) Hanser, Munich
24. Macosko, C. W., *Rheology: Principles, Measurements and Applications* (1994) Wiley-VCH, New York
25. Raju, V. R., Smith, G. G., Marin, G., Knox, J. R., Graessley, W. W., *J. Polym. Sci, Part B: Polym. Phys.* (1979) 17, p. 1183
26. Pearson, D. S., Ver Strate, G., von Meerwall, E., Schilling, F. C., *Macromolecules* (1987) 20, p. 1133
27. Mavridis, H., Shrorr, R., *J. Polym. Sci., Part A: Polym. Chem.* (1993) 49, p. 299
28. Wasserman, S. H., Graessley, W. W., *J. Rheol.* (1992) 36, p. 543
29. Inkson, N. J., McLeish, T. C. B., Groves, D. J., Harlen, O. G., *J. Rheol.* (1999) 43, p. 873
30. Anderssen, R. S., Mead, D. W., *J. Non-Newton. Fluid Mech.* (1998) 76, p. 299
31. Janzen, J., Colby, R. H., *J. Mol. Struct.* (1999) 569, p. 485
32. Fetters, J., Lohse D. J., Colby, R. H., In *Physical Properties of Polymers Handbook*. Mark, J. E. (Ed.) (1996) AIP Press, New York, p. 335

3 Melting Point of Polyethylene Resin

3.1 Introduction

The melting point of a polyethylene resin is its essential technical characteristics. The melting points are measured using the method of differential scanning calorimetry (DSC). Three parameters of a DSC melting curve, the position of the peak maximum (the melting point), the width of the melting transition, and the area under the melting curve (the heat of fusion) provide important information about the molecular structure of a resin, its crystallinity degree (see Chapter 4), and its structural uniformity. The value of the melting point gives the first qualitative information about the type of the resin and, to some degree, about mechanical properties that can be expected from the articles manufactured from the resin. This is particularly true for linear low density polyethylene (LLDPE) resins and very low density polyethylene (VLDPE) resins produced with single-site metallocene catalysts. All such resins are compositionally uniform ethylene/α-olefin copolymers and they all have strongly depressed melting points compared to the melting point of high density polyethylene (HDPE) resins [1–7]. On the other hand, melting points of compositionally nonuniform LLDPE resins produced with multi-site Ziegler-Natta catalysts and with chromium oxide catalysts are always relatively high (120 to 130 °C) and they only weakly depend on the average copolymer composition [7–15].

It is customary to measure the melting point and the heat of fusion of a polyethylene resin in a three-step procedure. First, a sample of a resin (5 to 10 mg) is packed into a small aluminum pan, placed into the oven of a DSC instrument, and heated to 140 or 150 °C at a high heating rate, for example 10 °C/min. Then the DSC oven with the melted sample inside is cooled to ~30 °C at a low rate, from 0.5 to 2 °C/min, to achieve orderly crystallization and thorough annealing of the resin sample. Finally, the crystallized/annealed sample is heated for the second time, its melting curve is recorded, and two parameters of the curve are measured, the temperature of the curve's maximum (the melting point T_m, °C) and the area under the curve (the heat of fusion ΔH_f, J/g). This second melting stage can be carried out at different heating rates, from 2 to 10 °C/min [16].

3.2 Melting Point of HDPE Resin

HDPE resins crystallize from the melt in a chain-folded lamellar morphology (see Section 1.9 in Chapter 1). A clear proof of the chain folding was originally discovered for individual linear high molecular weight paraffins starting from n-$C_{150}H_{302}$. A linear polymer segment of this size would correspond to an ethylene block M-(E)$_n$-M containing ~75 ethylene units [17]. The type of folding in paraffin crystals is schematically shown in Figure 3.1. The methyl end-groups of the paraffins and the folds are excluded from the crystal lattice; they are located at the surfaces of the lamellae (Figure 3.1). The folds connecting linear chain segments are tight, and the paraffin chains adjacently re-enter into the same crystal.

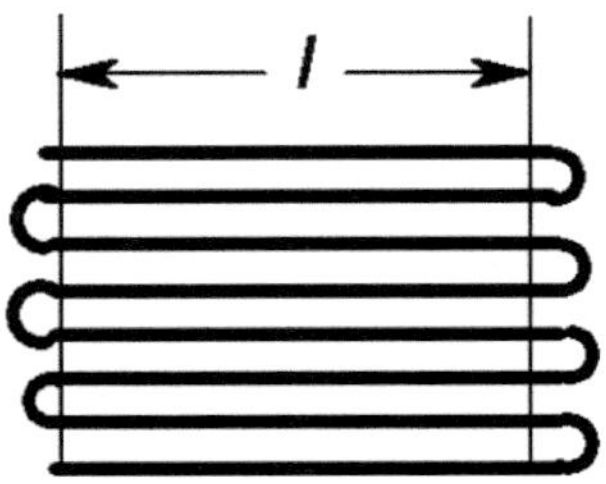

Figure 3.1
Scheme of paraffin and polyethylene lamellae;
l - lamella thickness

Polyethylene lamellae have a similar structure. They consist of orderly packed linear stretches of the same macromolecule or of several macromolecules [18–20]. The distance between neighboring chains in the lamellae is 4.4 to 4.6 Å. The bent segments of the polymer chains between the linear stretches (the folds) are called loops (Figure 1.16). The loops are usually relatively short and consist of a few ethylene units. The average length of the straight segment of a polymer chain between two loops is called the average lamella thickness, l_{av}; its size is usually given in the literature either as the physical length, in angstroms (Å), or as the average number of ethylene units in the segment, n_l. The n_l number in a lamella with the thickness of l_{av} (Å) is $n_l = l_{av}/2.54$, where 2.54 Å is the length of one ethylene unit in a polyethylene chain. The thickness of the lamellae of linear polyethylene strongly depends on the crystallization conditions, especially on the crystallization temperature. For example, when an HDPE resin is crystallized from the melt at a constant temperature of 120 °C, it initially produces lamellae with l_{av} ~220 Å (~90 ethylene units), but after the resin is kept for three hours at this temperature the lamellae thicken to ~330 Å (~130 ethylene units). If the same crystallization steps are carried out at 128 °C, the respective numbers are 260 Å (~100 ethylene units) and 480 Å (~190 ethylene units) [18].

This variability of lamella thickness is the principal reason why melting points of polyethylene resins are usually measured during the second melting stage. The complex three-step procedure, "melting → cooling (crystallization/annealing) → second melting", is used to erase the previous thermal history of the resin. The first rapid melting is followed by the stage of slow cooling/crystallization and annealing of the lamellae, and the melting curve is recorded after the lamellae are formed under standardized conditions. It is obvious also that crystallization of polyethylene under DSC conditions (slowly decreasing temperature) cannot result in the formation of lamellae with a strictly constant thickness. Rather, HDPE lamellae formed under these conditions have a certain distribution with respect to thickness (see Equation 3.2) and the l_{av} value represents their average thickness.

Our understanding of the melting behavior of polyethylene resins is based on detailed experimental data on the melting of crystalline paraffins. Paraffins with a carbon atom number lower than about n-C_{100} crystallize in the fully extended form. These crystals can be viewed as analogs of polyethylene lamellae but without the folds. The length of these crystals depends, of course, on the carbon atom number of the paraffin molecules. From the theoretical point of view, the dependence between the melting temperature (T_m) of a paraffin crystal and its crystal thickness l (or between the melting temperature of a polyethylene lamella and its thickness, l_{av}) is given by the Thompson-Gibbs equation [18, 21]:

$$T_m(\mathrm{K}) = T_m^o(1-\chi/l) \tag{3.1}$$

Here $T_m{}^o$ is the equilibrium melting temperature of the infinitely thick polyethylene crystal and χ is a constant related to the free energy of the fold surface, ~125 erg/cm^2, and to the bulk free energy of fusion, $2.8 \cdot 10^9$ erg/cm^3 [18]. Various estimations give the $T_m{}^o$ range from 419.2 K [18] to 414.6 K [21]. In the case of polyethylene crystals, the χ value in Equation 3.1 depends on the crystallization conditions, and various empirical expressions are usually used instead of Equation 3.1 to describe the dependence between the two experimentally measured parameters of polyethylene resins, T_m and l_{av}. Several such empirical expressions are described in the literature [16, 18, 22]; all of them provide quite close results.

It should be noted that when polyethylene resins of a very high molecular weight are crystallized from the melt at a very high pressure, the thickness of their lamellae greatly increases. At the limit, such chains can also crystallize in the fully extended form and produce very thick crystals with melting points in excess of 142 °C or 415 K [23, 24], which are similar to the predictions for the $T_m{}^o$ value for paraffins in Equation 3.1 [18, 21].

Modeling DSC curves of HDPE resins is based on two assumptions:

1. There exists an average thickness of the crystalline lamella, l_{av} (Å). Its value is determined by the thermodynamics of chain folding and crystallization kinetics. The l_{av} value for a given sample of linear polyethylene with a particular thermal history cannot be estimated solely from the theoretical viewpoint; it can only be

evaluated from the T_m value using Equation 3.1, provided that T_m is experimentally measured under isothermal conditions. The average number of ethylene units in such a lamella, n_{av}, is $l_{av}/2.54$.

2. The thickness of different polyethylene lamellae in a given resin is not strictly uniform. One can assume that it is distributed according to the Gauss law [16]:

$$Fr(l) = [\sigma \cdot \sqrt{2\pi}]^{-1} \cdot \mathrm{Exp}[-(l_{av} - l)^2/2\sigma^2] \tag{3.2}$$

where $Fr(l)$ is the fraction of lamellae with the thickness l (Å), l_{av} is the average lamella thickness, and σ is the width of the thickness distribution. This assumption is based on the fact that melting curves of linear polyethylene always have a significant width, from 6 to 8 °C, which is higher than the width of DSC melting curves of low molecular weight paraffin crystals.

Plotting Equation 3.2 in the coordinates "$Fr(l)$ as a function of l" gives a symmetrical Gauss curve with the maximum at l_{av} and the half-width σ. However, DSC melting curves are recorded as a continuously increasing temperature and are plotted in different coordinates: "heat flow ΔH (W/g) or heat capacity (J/K·mol) as a function of temperature." This peculiarity of the DSC coordinates requires a transformation of the Gauss function in the coordinates "ΔH as a function of T_m". This change of coordinates gives the following expression for the shape of the melting curve of linear polyethylene in the DSC coordinates [16]:

$$\begin{aligned} \Delta H = {} & T_m^o \cdot \chi \cdot (T_m^o - T_m)^{-2}/(\sigma \cdot \sqrt{2\pi}) \cdot \\ & \mathrm{Exp}\left\{-(T_m^o \cdot \chi)^2/2\sigma^2 \cdot (T_m^o - T_m^{av})^{-1} - (T_m^o - T_m)^{-1}\right]^2\right\} \\ & \text{as a function of } T_m \text{ (K)} \end{aligned} \tag{3.3}$$

Figure 3.2 compares two melting curves of linear polyethylene. The curve in Figure 3.2(a) is the experimental melting curve of an HDPE resin prepared with a single-site metallocene catalyst. The curve in Figure 3.2(b) is calculated with Equation 3.3 for T_m^{av} = 134.5 °C, using parameters of the Thompson-Gibbs equation from ref. [22]. The effective σ value in Equation 3.3 is the sum of two numbers: the instrumental broadening (which is specific for a particular DSC instrument and for particular recording conditions) and the real σ value in Equation 3.2 characterizing the distribution of lamella thickness. A comparison of the two curves in Figure 3.2 shows that Equation 3.3 provides an adequate model for the specific asymmetric shape of the melting curve of an HDPE resin.

Several features of DSC curves of HDPE resins require attention. Due to the nature of Equation 3.3, plotting the Gauss function in the DSC coordinates results in a shift of the curve's maximum. The maximums of melting curves calculated with Equation 3.3 are 1.0 to 1.5 °C higher than the T_m values used in the calculations.

This difference signifies that the real T_m^{av} values of HDPE resins, as well as the T_m^{av} values of medium density polyethylene (MDPE) resins with low α-olefin content, are slightly lower than the temperatures at the maximums of their DSC curves. The distortion of the Gauss function in the DSC coordinates described by Equation 3.3 (see Figure 3.2(b)) explains a noticeable difference in the shapes of the sides of DSC melting curves. The shape of such a curve at temperatures below the maximum reflects the existence of lamellae that are thinner than the average lamella, with $l < l_{av}$. On the other hand, the presence of thicker lamellae, with $l > l_{av}$, which melt at higher temperatures, remains virtually unnoticed in the DSC coordinates.

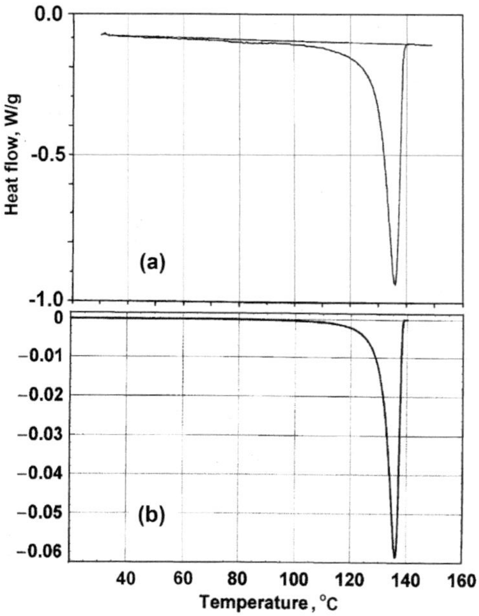

Figure 3.2
Modeling the melting DSC curve of HDPE resin: (a) experimental melting curve of linear polyethylene produced with single-site catalyst; (b) calculation with Equation 3.3

In the literature, melting points of low molecular weight organic crystals or inorganic substances are measured as an intercept between the tangent to the low-temperature side of the melting curve and its baseline. However, the existence of a distribution in the lamella thickness of semi-crystalline polymers (Equation 3.2) does not justify this practice in the case of polyethylenes. The shape of the low-temperature side of the DSC melting curve of an HDPE resin is mostly determined by melting of thinner lamellae, and the positioning of the tangent to such a curve is arbitrary.

3.3 DSC Melting Curves and Melting Points of LLDPE and VLDPE Resins Produced with Single-Site Catalysts

3.3.1 Effect of Crystallization Conditions

Compositionally uniform ethylene/α-olefin copolymers prepared with single-site metallocene catalysts represent a convenient model to examine the effect of crystallization conditions on the shape of DSC melting curves. Figure 3.3 demonstrates this effect using as an example two ethylene/1-hexene copolymers prepared with different metallocene catalysts in toluene slurry at 70 °C [25]. The copolymers have a different composition: the sample in Figure 3.3(A) contains 0.8 mol % of 1-hexene and the sample in Figure 3.3(B) 2.4 mol %. Their crystallinity degree is also different, 63% and 42%, respectively. Ethylene copolymerization reactions in toluene slurry are accompanied by orderly crystallization of polymer molecules. DSC peaks of both nascent copolymers (materials recovered from polymerization reactors without any additional thermal treatment) are relatively narrow (Figure 3.3(A) and Figure 3.3(B)). When the mixture of the two copolymers is slowly crystallized from toluene solution, each component forms separate crystals (Figure 3.3(C)).

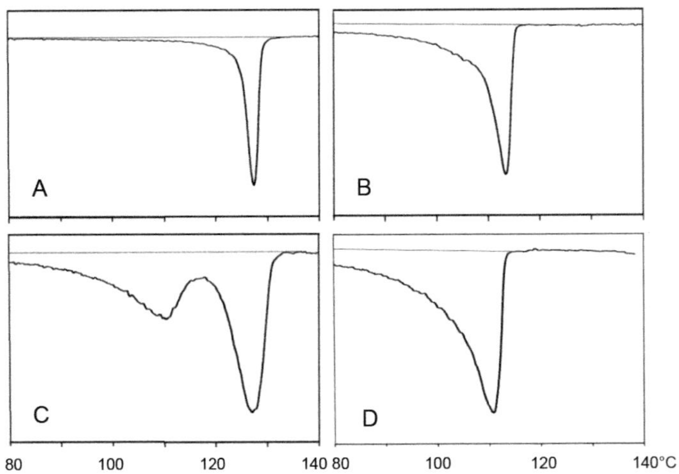

Figure 3.3 DSC melting curves in the narrow temperature range 80 – 140 °C: A and B – nascent copolymers with C_M^{copol} = 0.8 and 2.4 mol %, respectively, produced with two metallocene catalysts; C – mixture of the two copolymers crystallized from solution; D – re-melted copolymer B

When these compositionally uniform copolymers are crystallized from the melt, the lamella-formation process is much faster and less orderly than during crystallization from solution or during a polymerization reaction. The DSC peak of a thermally pretreated copolymer (Figure 3.3(D)) is broader than the DSC peak of the solution-crystallized copolymer (Figure 3.3(B)), and its melting peak is positioned at a lower temperature. This comparison emphasized the need for a thoroughly followed protocol in the preparation of resin samples for the DSC analysis. Analytical laboratories often examine polymer samples with unknown or poorly defined thermal history. A standardized procedure, for example, pre-melting of a resin specimen in a hot press, is a desired step for a DSC sample preparation.

3.3.2 Effect of Copolymer Composition

Figure 3.4 shows melting points of compositionally uniform ethylene/1-hexene copolymers produced with single-site metallocene catalysts as a function of their composition. The copolymers contain from 0.5 to 3.5 mol % of 1-hexene. All the materials were thoroughly homogenized before the DSC analysis by hot-pressing polymer powders into thick film. The melting points were measured during the second melting step; each corresponds to the maximum on the DSC melting curve. The results show that the melting point decreases with an increase in the 1-hexene content in the copolymers. This section describes a DSC model which provides an explanation for the steep dependence between T_m and C_M^{copol} values evident from Figure 3.4.

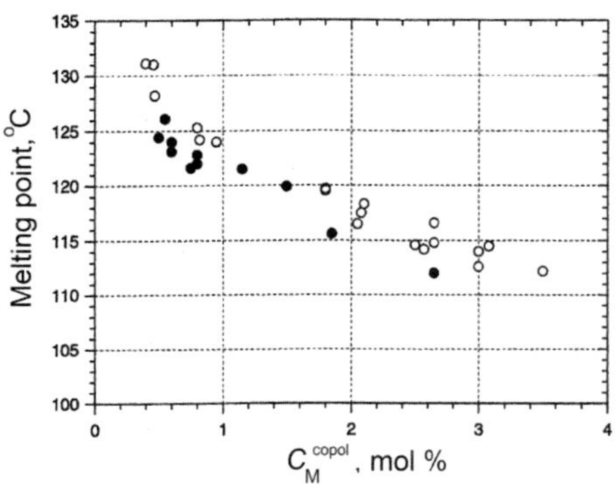

Figure 3.4
Melting temperatures of ethylene/1-hexene copolymers of different composition produced with single-site metallocene catalyst; heating rates are 10 °C/min (●) and 2 °C/min (○)

Basic statistical expressions used for modeling the melting behavior of compositionally uniform ethylene/α-olefin copolymers are given in Section 1.7 of Chapter 1. For the goal of the DSC modeling, a copolymer chain can be viewed as consisting of a set of monomer sequences; blocks of ethylene units, M–$(E)_n$–M, and blocks of α-olefin units, E–$(M)_m$–E. All commercial LLDPE and VLDPE resins contain relatively small molar fractions of α-olefins; their C_M^{copol} values vary from 2 to 10 mol % and most of their α-olefin blocks $E-(M)_m-E$ are very short, with m from 1 to 3. Crystallinity of all these materials, and therefore all their DSC features, are entirely determined by the presence of long ethylene blocks $M-(E)_n-M$ in the copolymer chains.

Section 1.7 in Chapter 1 gives two statistical functions that describe the distribution of ethylene units in blocks of different size. The first statistical function is $\delta(E)_n$, the fraction of ethylene units in $M-(E)_n-M$ blocks containing n monomer units normalized to the total molar content of ethylene units in the copolymer (Equation 1.17). The second statistical function is $\Sigma(E)_n$, the fraction of ethylene units in the sum of all long ethylene blocks starting with a block containing n monomer units, that is, the fraction of ethylene units in the sum of blocks $M-(E)_n-M$, $M-(E)_{n+1}-M$, $M-(E)_{n+2}-M$, and so on (Equation 1.18). Both these expressions contain a single variable, the molar percent of α-olefin in the copolymer, C_M^{copol}. Two examples of the $\delta(E)_n$ function represented by Equation 1.17 are shown in Figure 1.10: one for a copolymer with C_M^{copol} = 3 mol %, which is a typical composition of an LLDPE resin, and another for a copolymer with C_M^{copol} = 8 mol %, a typical composition of a VLDPE resin. A comparison of the two curves in Figure 1.10 demonstrates that the lower the α-olefin content in a copolymer, the higher is the fraction of ethylene units in longer (crystallizable) blocks.

As Figure 1.10 shows, every distribution of ethylene units in blocks $M-(E)_n-M$ has a maximum which corresponds to the size of the most abundant ethylene sequence. The position of this maximum is a function of copolymer composition. The maximum of the $\delta(E)_n$ function is positioned at $d[\delta(E)_n]/dn = 0$. Differentiation of Equation 1.17 gives the position of the maximum:

$$n_{max} = -[\ln(1 - C_M^{copol}/100)]^{-1} \tag{3.4}$$

Figure 3.5 shows the plot of the size of the most abundant ethylene block in an ethylene/α-olefin copolymer, n_{max}, as a function of the α-olefin content, C_M^{copol}. As expected, the n_{max} value changes from a very large number, ~ 200 ethylene units, at C_M^{copol} = 0.5 mol % (composition of a typical MDPE resin), to a quite small number, ~ 10 ethylene units, at C_M^{copol} = 10 mol % (composition of a VLDPE resin).

Figure 3.6 replots melting temperatures of ethylene/1-hexene copolymers prepared with a single-site metallocene catalyst (from Figure 3.4) in the coordinates "T_m as a function of n_{max}" with n_{max} values calculated from C_M^{copol} using Equation 3.4. This plot has two distinct ranges. When the copolymers have a low 1-hexene content (C_M^{copol} from 0.8 to 0.3 mol %) and large n_{max} values, from ~ 120 to 450, the melting temperature changes with n_{max} very little, in the 120 to 125 °C range. When the copolymers have a high 1-hexene content (C_M^{copol} higher than 0.8 mol %) and small n_{max} values, below ~ 120, the melting temperature sharply decreases as the n_{max} value decreases.

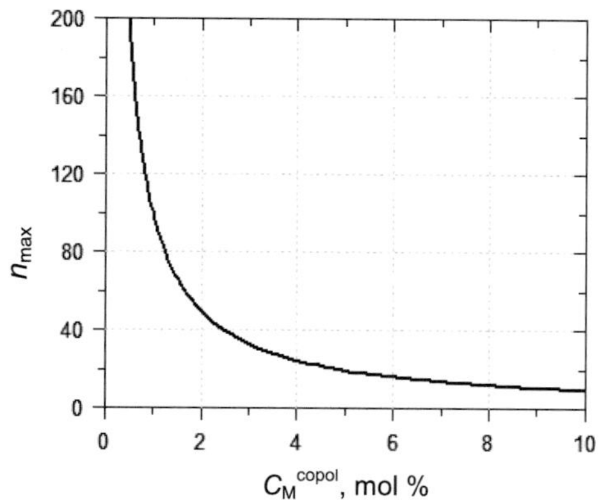

Figure 3.5
The size of the most abundant ethylene block (n_{max}) as a function of α-olefin content in ethylene/α-olefin copolymers (Equation 3.4)

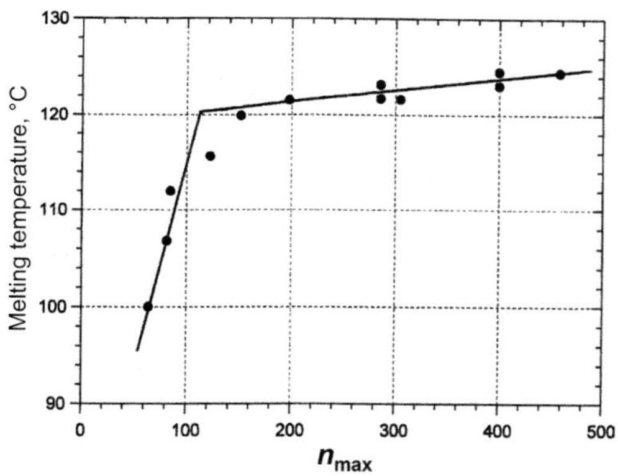

Figure 3.6
Melting points of ethylene / 1-hexene copolymers prepared with single-site metallocene catalyst (from Figure 3.4, 10 °C/min heating rate) as a function of the size of the most abundant ethylene block, n_{max}

The first part of this correlation (the flat range of the $T_m - n_{max}$ dependence) can be rationalized in view of the chain folding process during crystallization of polyethylene chains, which is described in the previous section. In the case of ethylene copolymers, the situation is complicated by several circumstances [18–20]:

1. The copolymers have ethylene units arranged in a variety of blocks, from very long to very short (Figure 1.10).
2. The nature of the loops that connect straight segments of the ethylene blocks in the lamellae (Figure 1.16 and Figure 3.1) is not uniform; some loops include α-olefin units (most of them isolated α-olefin units, E-M-E sequences), other loops are parts of folded ethylene blocks.
3. The average thickness of the lamellae depends on the rate of crystallization from the melt, the same phenomenon that occurs during crystallization of HDPE resins with linear chains.

To avoid the latter problems, the modeling of DSC melting curves of compositionally uniform LLDPE and VLDPE resins described below concentrates on the second part of the plot in Figure 3.6, the steep dependence between T_m and n_{max}. In this T_m range, when the n_{max} values are low, the dominant factor determining the melting temperature is the average size n of crystallizable ethylene blocks M–(E)$_n$–M, rather than chain folding of long ethylene blocks or lamella thickening during crystallization.

3.3.3 Crystallization Process of Compositionally Uniform Ethylene/α-Olefin Copolymers. Model for Primary Crystallization

The principal assumption of the DSC model for compositionally uniform LLDPE resins is described in Table 3.1. It lists the sequence of crystallization steps which take place when ethylene blocks M–(E)$_n$–M of different size crystallize from the melt at a gradually decreasing temperature.

Table 3.1 Crystallization of Different Ethylene Blocks M-(E)$_n$-M in Compositionally Uniform Ethylene/α-Olefin Copolymers

Type of M–(E)$_n$–M block	Block length and content of ethylene units	Type of crystallization and DSC curve shape
Long crystallizable blocks	$n > n_{max}$ (Equation 3.4), ethylene unit content, $\Sigma(E)_n$, is given by Equation 1.18	Primary crystallization at higher temperatures; the shape of DSC curve is given by Equation 3.3
Short crystallizable blocks	$n_{min} > n > n_{max}$, ethylene unit content in each (E)$_n$ block is given by Equation 1.17	Secondary crystallization at lower temperatures; the shape of DSC curve is given by Equation 3.5
Short non-crystallizable blocks, $n \le n_{min}$	n_{min} = 8 to 9, ethylene unit content is calculated as $1 - \Sigma(E)_{n,min}$ (Equation 1.18)	Blocks are too short to crystallize at 20–30 °C

In the first step of the crystallization process, at relatively high temperatures, all long ethylene blocks with n higher than a particular limit value, n_{limit}, rapidly crystallize in the chain-folded morphology and provide the framework for the spherulite structure. One can assume that the n_{limit} value is slightly higher than the number of ethylene units in the most abundant ethylene sequence, n_{max}, which is defined by Equation 3.4: $n_{limit} = k \cdot n_{max}$ where $k \sim 1.1$ to 1.2. When the crystallized copolymer is subsequently melted, the T_m^{av} value for these thickest lamellae consisting of long ethylene blocks is lower than the T_m value for linear polyethylene because the l_{limit} value for the thickest lamellae in the copolymers ($l_{limit} = 2.54 \cdot n_{limit}$) is smaller than for linear polyethylene.

Melting curves of the thick lamellae formed during the primary crystallization step are described by the same DSC model as the one used in Section 3.2 to describe melting of HDPE resins. Similarly to the principal assumption of the model in Section 3.2, the number of ethylene units in the thick lamellae, n_{limit}, is not uniform but is distributed according to the Gauss function in Equation 3.2. When these thick lamellae melt during a DSC experiment, their melting curve is described by Equation 3.3 (the Gauss function in the DSC coordinates) but with the maximum at a lower T_m^{av}, which is calculated with Equation 3.1 for the particular n_{max} value. Special attention should be paid when these T_m^{av} values for ethylene copolymers are calculated. As Figure 3.5 shows, the dependence between T_m and n_{max} in the range of the plot at $n_{max} < 150$ is quite steep, and even a small error in the measurement of the C_M^{copol} value (which is carried out either by the infrared or the nuclear magnetic resonance method) can strongly affect the estimation of both the n_{max} value and the calculated T_m value.

3.3.4 Model for Secondary Crystallization

The second stage of the three-step DSC analysis, the stage of crystallization/annealing, is usually carried out at a low cooling rate. After the primary crystallization process of long ethylene blocks with $n \geq n_{limit}$ is completed, the temperature of the resin sample continues to decrease and the secondary crystallization process, crystallization of the remaining shorter M–(E)$_n$–M blocks, begins (Table 3.1). This is a gradual and a much slower process. Such two-stage crystallization processes and their effect on experimental DSC data for melt-crystallized and solution-crystallized polyethylene resins have been thoroughly investigated [22]. Even when the sample is cooled to room temperature, very short ethylene blocks M–(E)$_n$–M with n lower than a particular minimum n value, n_{min} ~ 8 to 10, still do not crystallize, although even these shortest blocks can be forced to crystallize at lower temperatures.

The model for the secondary crystallization of short ethylene blocks is suitable only for ethylene copolymers with C_M^{copol} values higher than ~ 1 mol %, which is in the range of the plot in Figure 3.6 where the T_m value strongly depends on n_{max}. The model is based on the following assumptions:

1. Most of the shorter ethylene sequences crystallize in the fully extended form, like short paraffin molecules. Indeed, the typical n_{max} values in this range vary from ~ 50 to 120 (Figure 3.6), very similar to the minimum equilibrium lamella size for linear polyethylene, 90 to 100 ethylene units. Surfaces of these thin lamellae do not contain tightly folded segments of linear chains, as in Figure 3.1; they consist of branching points in the copolymer chains, which are mostly single α-olefin units, E–M–E sequences, and the blocks containing two α-olefin units, E–M–M–E, as well as the end-groups of polymer chains.
2. If a resin sample is crystallized slowly, one can assume that each ethylene block with the n value in the range between n_{limit} and n_{min} co-crystallizes mostly with

ethylene blocks of approximately the same size and that they form lamellae of a relatively uniform small thickness. The possibility of such an orderly crystallization depends on the relative rates of two processes: the rate of polyethylene crystallization, which is quite high [20], and the cooling rate during the crystallization stage, which is preferably kept low. The fraction of ethylene units in the M–(E)$_n$–M block, the $\delta(\mathrm{E})_n$ value in Equation 1.17, gives the fraction of these units in lamellae with the thickness of 2.54·n (Å).

The dependence between $\delta(\mathrm{E})_n$ and n is represented in a graphic form by the plots similar to those in Figure 1.10. As discussed above, DSC melting curves are plotted in the coordinates "heat flow ΔH as a function of temperature". If one assumes that the heat of fusion per one ethylene unit does not depend on the lamella thickness, the second of the above assumptions signifies that the heat flow is proportional to the $\delta(\mathrm{E})_n$ value given by Equation 1.17. To produce a theoretical DSC melting curve for the thin lamellae formed during the secondary crystallization process, one has to change the coordinates in Figure 1.10 from "$\delta(\mathrm{E})_n$ as a function of n" to "ΔH as a function of T_m," the same transformation as in the crystallization model for HDPE resins described in Section 3.2. The final expression for the shape of the DSC melting curve of secondary crystals is [16]:

$$\Delta H = (C_\mathrm{M}^\mathrm{copol}/100)^2 \cdot \left[T_\mathrm{m}^\mathrm{o} \cdot (\chi/2.54)\right]^2 \cdot (T_\mathrm{m}^\mathrm{o} - T_\mathrm{m})^{-3} \cdot (1 - C_\mathrm{M}^\mathrm{copol}/100)^\theta \qquad (3.5)$$

where $\theta = T_\mathrm{m}^\mathrm{o} \cdot (\chi/2.54)/(T_\mathrm{m}^\mathrm{o} - T_\mathrm{m}) - 1$, as a function of T_m (K).

3.3.5 Combined DSC Model for Compositionally Uniform LLDPE and VLDPE Resins

The combination of Equation 3.3 and Equation 3.5 describes the overall shape of the DSC melting curve of a compositionally uniform ethylene/α-olefin copolymer. Figure 3.7(a) shows the experimental DSC melting curve of an ethylene/1-hexene copolymer with $C_\mathrm{M}^\mathrm{copol}$ = 3.2 mol % prepared with a single-site metallocene catalyst. Figure 3.7(b) shows the calculated DSC curve using a combination of Equation 3.3 and Equation 3.5. Curve 1 in Figure 3.7(b) is calculated with Equation 3.3; it represents the melting of thick lamellae formed during the primary crystallization step. The curve is similar shape to the melting curve of the HDPE resin prepared with the same catalyst (Figure 3.2(b)) although the Tm value of the copolymer is significantly lower. Curve 2 in Figure 3.7(b) represents the melting of thin lamellae formed during the secondary crystallization process; it is calculated with Equation 3.5. The areas under the two parts of the calculated DSC curve correspond to the fractions of crystallizable material in the two types of lamellae formed during the primary and the secondary crystallization processes; 68 and 32%, respectively.

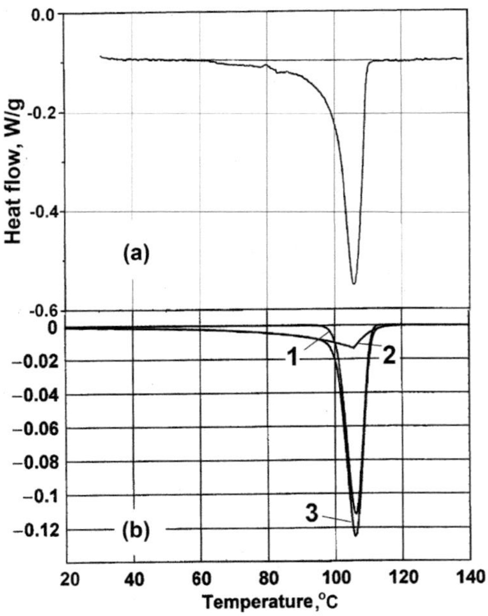

Figure 3.7
Modeling DSC melting curves of compositionally uniform ethylene / α-olefin copolymers. (a) Experimental DSC curve of ethylene / 1-hexene copolymer with C_M^{copol} = 3.2 mol % prepared with single-site metallocene catalyst; (b) calculated DSC melting curves: 1 – melting curve of primary lamellae (calculated with Equation 3.3); 2 – sum of melting curves of secondary lamellae (calculated with Equation 3.5); 3 – combined calculated DSC curve [16].

Figure 3.8 gives another example of using the combined DSC model for compositionally uniform ethylene/α-olefin copolymers, an ethylene/1-hexene VLDPE plastomer containing 7 mol % of 1-hexene. This material was also prepared with a single-site metallocene catalyst. A comparison of the two plots in Figure 3.8 confirms that the combined model gives realistic results for ethylene plastomers as well, both in terms of the shape of the DSC melting curve and the position of the peak maximum.

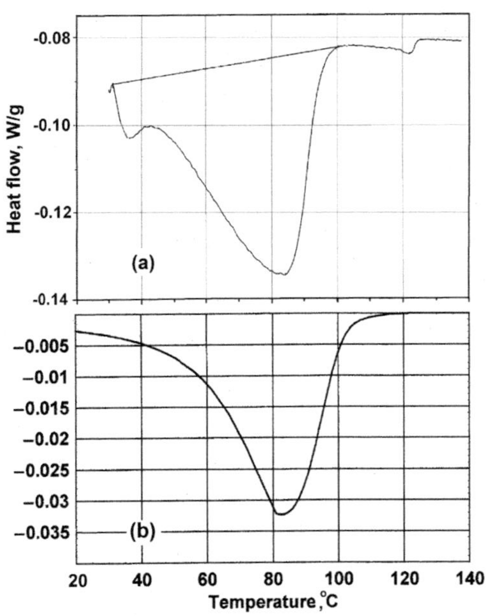

Figure 3.8
Modeling DSC melting curves of ethylene / α-olefin plastomers: (a) experimental DSC curve of ethylene / 1-hexene copolymer with C_M^{copol} = 7 mol % prepared with single-site catalyst; (b) modeled DSC curve calculated with Equation 3.3 and Equation 3.5 [16]

The same approach can be applied to modeling DSC melting curves of other compositionally uniform ethylene copolymers of this type. It should be stressed that the presented model is not intended for a precise reproduction of the experimental DSC data. Rather, it provides an interpretation of the experimental data in terms of the two-stage crystallization process described in Table 3.1.

3.4 DSC Melting Curves and Melting Points of LLDPE Resins Produced with Multi-Site Catalysts

All ethylene/α-olefin copolymers produced with supported titanium-based Ziegler-Natta catalysts, that is, the overwhelming majority of commodity LLDPE resins, as well as LDLPE resins produced with chromium oxide catalysts, are, in reality, complex mixtures each consisting of several copolymer components (see discussion in Section 1.8 of Chapter 1). The components in each mixture have widely different compositions and molecular weights; several examples are shown in Table 1.7 and Table 1.8, Figure 1.12 and Figure 1.13. Each such copolymer component is produced by a particular population of active centers in the catalyst. The number of the copolymer components ranges from four to six and the α-olefin content in them varies from 0.3 to over 15 mol % [7, 8].

Supported Ziegler-Natta catalysts consist of spherical inorganic particles (silica or magnesium dichloride) with a diameter of 20 to 100 micron. The particles are porous and filled with catalytically active ingredients. Each catalyst particle contains all the types of active centers in close proximity and each active center produces thousands of macromolecules in the course of a copolymerization reaction. As a result, copolymer molecules of different composition remain mixed and packaged within a single polymer particle [7, 8]. Most LLDPE resins are pelletized immediately after exiting a polymerization reaction. This procedure leads to a further homogenization of the resin components. Modeling DSC melting curves of such complex, thoroughly mixed copolymer mixtures represents a significant challenge.

As described in the introduction to this chapter, DSC analysis of ethylene/α-olefin resins consists of three stages: the first melting stage, a slow cooling (crystallization/annealing) stage, and the second melting stage. The crystallization stage of the multicomponent mixtures can be described by the same sequence of steps as that shown in Table 3.1: the primary crystallization of long ethylene blocks followed by the secondary crystallization of short ethylene blocks. When the mixture of different copolymer components is melted in a DSC oven, it leads to its additional homogenization. When the temperature of such a multi-component melt is slowly decreased, the longest ethylene blocks in any of the copolymer components (but mostly the long blocks in the components with the lowest α-olefin content) crystallize together at higher temperatures and produce the thickest lamellae. Shorter ethylene sequences in all the copolymer components also co-crystallize together at lower temperatures.

The simplest approach to modeling the melting process of such copolymer mixtures is similar to modeling of compositionally uniform copolymers described in Section 3.3. Figure 3.9(a) shows the experimental melting curve of an LLDPE resin produced with a supported Ziegler-Natta catalyst. The average hexene content in the copolymer is 3.6 mol %. The catalyst contains five different types of active centers. Table 3.2 gives the fraction of the copolymer component produced by each type of active center and the 1-hexene content in each fraction. Copolymer components IV and V have the lowest 1-hexene content, about 1 mol %, and their chains mostly consist of long ethylene blocks. Thick lamellae formed from these ethylene blocks melt at a relatively high temperature; they are represented in Figure 3.9(a) by a narrow melting peak with a maximum at ~128 °C. This T_m value, which is usually cited as "the melting point of an LLDPE resin," is 17 °C higher than the melting point of a compositionally uniform resin with the same C_M^{copol} value (compare to Figure 3.7). Thinner lamellae formed during crystallization of shorter ethylene blocks in copolymer components IV and V and in most of component III (the largest component in the mixture) melt at lower temperatures; they produce a characteristic low-temperature "tail" on the DSC curve in Figure 3.9(a).

Table 3.2 Parameters Used in Modeling DSC Melting Curve of Multi-Component Ethylene/1-Hexene Copolymer with Average 1-Hexene Content 3.6 mol %

Copolymer component:	I	II	III	IV	V
Fraction, %:	5	20	44	23	8
1-Hexene content (C_M^{copol}), mol %:	15	6.5	4.5	1.3	0.9

To model melting curves of such complex mixtures, all crystalline lamellae in them are separated into two populations. The population containing the thickest lamellae is formed at high temperatures during the primary crystallization process from relatively long ethylene blocks. These long ethylene blocks are present mostly in macromolecules belonging to components IV and V (Table 3.2) but, to some degree, also in component III, although in a much smaller fraction. The component IV is the largest highly crystalline component in this mixture, and one can assume that the thickness of the largest number of lamellae is determined by the number of ethylene units in the most abundant blocks in this component, n_{max}(IV). This n_{max}(IV) value is calculated with Equation 3.4. According to both DSC models described in Section 3.2 and Section 3.3, the thickness of these lamellae is not uniform but is distributed according to the Gauss function distorted in the DSC coordinates (Equation 3.3).

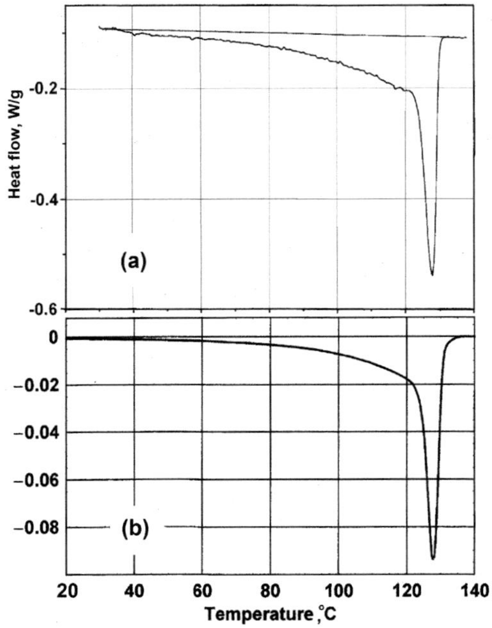

Figure 3.9
Modeling DSC melting curve of LLDPE resin consisting of several components: (a) experimental DSC curve of ethylene/1-hexene copolymer with C_M^{copol} = 3.6 mol % prepared with supported Ziegler-Natta catalyst; (b) modeled DSC curve for combination of components in Table 3.2

After these long ethylene blocks form thick primary lamellae and while the temperature of the resin sample continues to decrease, the secondary crystallization process starts and the population of thinner lamellae begins to form. The secondary crystallization process takes place over a wide temperature range, ~80 °C. Equation 3.5 describes the shape of melting curves of the secondary lamellae in the DSC coordinates. Short blocks M–(E)$_n$–M with $n < n_{max}$(IV) in each of Components V, IV, III, and II contribute to this part of the melting curve in proportion to the fraction of ethylene units in these blocks and the amount of the copolymer component (Table 3.2).

Thus, the experimental DSC melting curve of such a multi-component mixture can be viewed as two overlapping curves: one a Gauss curve in the DSC coordinates (Equation 3.3) representing melting of all the thickest lamellae and another curve an asymmetric "tail" representing melting of a large number of various thinner lamellae (Equation 3.5). The final modeling curve calculated for the five-component copolymer mixture in Table 3.2 is shown in Figure 3.9(b). The high-temperature peak represents melting of thick lamellae (l ~ 280 Å) that mostly come from copolymer components IV and V. The low-temperature tail represents melting of thinner lamellae, which mostly come from copolymer chains in components III (~60%) and II (~20%), as well as the shortest ethylene blocks in components IV and V (~20%). As to the copolymer component I in this resin, it has a very high 1-hexene content and it is practically amorphous.

Although the approach to modeling DSC melting points of two types of ethylene/α-olefin copolymers, compositionally uniform LLDPE resins and compositionally nonuniform LLDPE resins produced with Ziegler-Natta catalysts, is the same, it gives very different results. In the case of compositionally uniform copolymers, the α-olefin content is approximately the same for all macromolecules in the resin. As a result, the size of the most abundant ethylene block n_{max} steeply decreases as the C_M^{copol} value increases (Equation 3.4), and, correspondingly, the effective melting point of such resins rapidly decreases with the C_M^{copol} value (Figure 3.3). On the other hand, sharp melting peaks of compositionally nonuniform resins (Figure 3.9) are produced by melting of the thickest lamellae from copolymer components with the lowest α-olefin content (for example, components IV and V in Table 3.2). Active centers in Ziegler-Natta catalysts that produce these materials copolymerize α-olefins with ethylene very poorly [7, 8]; the comonomer content in these components is always low, and as a result, effective T_m values of such resins are always high.

To illustrate this difference, Table 3.3 compares estimated melting points (maximums on DSC curves) for two series of copolymers with average C_M^{copol} values ranging from 0.5 to 4 mol %. The second row in the table gives the calculated T_m values for compositionally uniform copolymers. The third and the fourth rows describe compositionally nonuniform copolymers, the average composition of the sum of copolymer components IV and V and their melting point.

Table 3.3 Melting Points of Two Sets of Ethylene/α-Olefin Copolymers

Average copolymer composition, $(C_M^{copol})_{av}$, mol %:	0.5	1.0	2.0	3.0	4.0
Melting point of compositionally uniform copolymer with given C_M^{copol}, °C:	130	126	118	113	108
Averaged composition of components IV + V in compositionally nonuniform copolymer, mol %:	0.2	0.4	0.6	0.8	0.9
Melting point of components IV + V in compositionally nonuniform copolymer, °C:	132	130	128	127	126

Two conclusions follow from this comparison:

1. The contents of α-olefins in the sums of components IV and V in compositionally nonuniform copolymers are always much lower than the average $(C_M^{copol})_{av}$ values for the whole copolymers. The effective melting points of compositionally nonuniform LLDPE resins reflect the melting behavior of copolymer components with the lowest comonomer content. These melting points are always significantly higher than the melting points of compositionally uniform LLDPE resins of the same copolymer composition.

2. As the average α-olefin content in a compositionally nonuniform copolymer increases, the averaged C_M^{copol} values for the sum of components IV and V also increase, but they always remain relatively low, 0.2 to 0.9 mol % in the examples in Table 3.3. The data in Figure 3.6 show that this range of copolymer compositions corresponds to the flat part of the T_m vs. n_{max} curve, where the T_m value changes relatively little with n_{max} (and with C_M^{copol}). As a result, melting points of compositionally nonuniform LLDPE resins decrease relatively little as their average content of α-olefin increases.

3.5 DSC Melting Curves of LLDPE Resins Produced with Post-Metallocene Catalysts

The class of post-metallocene polymerization catalysts is very diverse; it includes numerous complexes of early- and late-transition metals with multi-dentate ligands [26–30]. Most post-metallocene catalysts are homogeneous (soluble) catalysts, similarly to metallocene catalysts. However, they often contain several types of active centers which differ in their ability to copolymerize α-olefins with ethylene. When ethylene/α-olefin copolymerization reactions with these catalysts are carried out in solution, each active center acts independently of other active centers; it produces a large number of macromolecules and forms a polymer micro-domain surrounding the center. As a consequence, ethylene/α-olefin copolymers produced with post-metallocene catalysts are not compositionally uniform; they contain from 2 to 4 components with different α-olefin content. If such polymer mixtures are not thoroughly homogenized before DSC analysis, each component crystallizes separately and melts at a different temperature [25].

The DSC melting curve of such a resin is an overlap of melting peaks of several individual components. The shape of each curve is described by the Gauss function distorted in the DSC coordinates (Equation 3.3) and the T_m^{av} value is calculated with Equation 3.1 for $l = 2.54 \cdot n_{limit}$, where $n_{limit} = k \cdot n_{max}$ ($k \sim 1.1$ to 1.2). In turn, the n_{max} value is calculated with Equation 3.4 as a function of the composition C_M^{copol} of a given copolymer component in the mixture.

As an example, Figure 3.10 shows the DSC melting curve of a VLDPE resin, an ethylene/1-decene copolymer prepared with a post-metallocene catalyst; its average 1-decene content is 5 mol % [31]. This product is a mixture of four components. Their compositions and relative contents are listed in Table 3.4. The material is mostly amorphous, its crystallinity degree is merely ~15%, and the two largest components in the mixture, components I and II, both have a high 1-decene content. As Figure 3.10 demonstrates, DSC analysis over-emphasizes the presence of relatively minor but highly crystalline Components III and IV in the copolymer mixture, which both have a low 1-decene content, 2 to 2.5 mol %. After copolymer mixtures of this type are thoroughly homogenized in the melt, their DSC melting curves become similar to that in Figure 3.9(a).

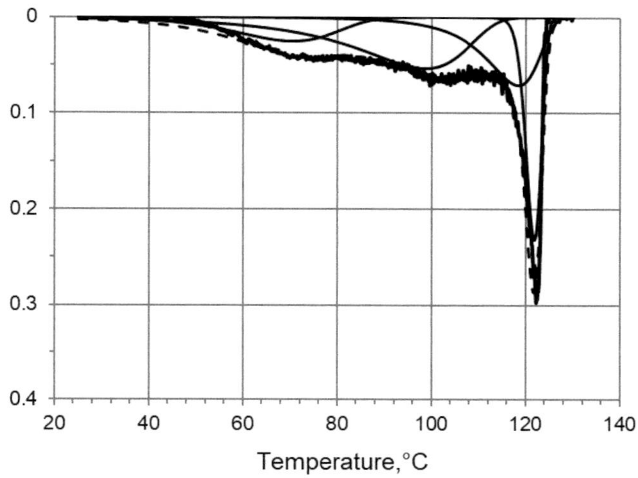

Figure 3.10 DSC melting curve of ethylene/1-decene copolymer (C_M^{copol} = 6.5 mol %) prepared with a post-metallocene catalyst (points) and its resolution into individual components.

Table 3.4 Components of Ethylene/1-Decene Copolymer Prepared with Post-Metallocene Catalyst; 1-Decene Content 6.5 mol % [31]

Copolymer component:	I	II	III	IV
Component fraction, %:	48	43	11	8
1-Decene content (C_M^{copol}), mol %:	~ 8	4.9	2.5	1.9
Melting point (T_m), °C:	~ 80	99	118	122

References

1. Bortolussi, F., Broyer, J.-P., Spitz, R., Boisson, C., *Macromol. Chem. Phys.* (2002) 203, p. 2501
2. Piel, C., Karssenberg, F. G., Kaminsky, W., Mathot, V. B. F., *Macromolecules* (2005) 38, p. 6789
3. Koivumaki, J., *Polym. Bull.* (1996) 36, p. 12
4. Bensason, S., Minick, J., Moet, A., Chum, S., Hiltner, A., Baer, E., *J. Polym. Sci., Part B: Polym. Phys.* (1996) 34, p. 1301
5. Aitola, E., Puranen, A., Setälä, H., Lipponen, S., Leskelä, M., Repo, T., *J. Polym. Sci., Part A: Polym. Chem.* (2006) 44, p. 6569
6. Nomura, K., Itagaki, K., Fujiki, M., *Macromolecules* (2005) 38, p. 2053
7. Kissin, Y. V., *Alkene Polymerization Reactions with Transition Metal Catalysts* (2008) Elsevier, Amsterdam, Chapters 2, 3

8. Krentsel, B.A., Kissin, Y.V., Kleiner, V.I., Stotskaya, S.S., *Polymers and Copolymers of Higher α-Olefins* (1997) Hanser, Munich, Chapter 8
9. Longo, P., Grassi, A., Oliva, L., *Makromol. Chem.* (1990) 191, p. 2387
10. Xu, G.X., Lin, S.A., *Macromolecules* (1997) 30, p. 685
11. Oliva, L., Immirzi, A., Tedesco, C., Venditto, V., Proto, A., *Macromolecules* (1999) 32, p. 2675
12. Naga, N., Mizunuma, K., Sadatoshi, H., Kakugo, M., *Macromolecules* (1997) 30, p. 2197
13. Palza, H., Lopez-Majada, M., Quijada, R., Benavente, R., Perez, E., Cerrada, M.L., *Macromol. Chem. Phys.* (2005) 206, p. 1221
14. Rulhoff, S., Kaminsky, W., *Macromol. Chem. Phys.* (2006) 207, p. 1450
15. De Rosa, C., Auriemma, F., *Macromolecules* (2006) 39, p. 249
16. Kissin, Y.V., *J. Polym. Sci., Part B: Polym. Phys.* (2011) 49, p. 195
17. Ungar, G., Stejny, J., Keller, A., Bidd, I., *Science* (1985) 229, p. 386
18. Chivers, R.A., Barham, P.J., Martinez-Salazar, D., Keller, A., *J. Polym. Sci., Part B: Polym. Phys.* (1982) 20, p. 1717
19. Barham, P.J., Chivers, R.A., Jarvis, D.A., Martinez-Salazar, J., Keller, A., *J. Polym. Sci., Polym. Lett.* (1981) 19, p. 539
20. Barham, P.J., Jarvis, D.A., Keller, A., *J. Polym. Sci., Part B: Polym. Phys.* (1982) 20, p. 1733
21. Wunderlich, B., *Macromolecular Physics*, vol. 3 (Crystal Melting) (1980) Academic Press, New York
22. Mirabella, F.M., *J. Polym. Sci., Part B: Polym. Phys.* (2001) 39, p. 2800
23. Bassett, D.C., *Principles of Polymer Morphology* (1981) Cambridge University Press, Cambridge, Chapter 7
24. Wunderlich, B., Miello, L., *Makromol. Chem.* (1968) 118, p. 2503
25. Kissin, Y. V., *J. Macromol. Sci., Part A: Pure Appl. Chem.* (2017) 54, p. 6
26. Gromada, J., Carpentier, J.-F., Mortreux, A., *Coord. Chem. Rev.* (2004) 248, p. 397
27. Domski, G. J., Rose, J. M., Coates, G. W., Bolig, A. D., Brookhart, M., *Prog. Polym. Sci.* (2007) 32, p. 30
28. Chum, P. S., Swogger, K. W., *Prog. Polym. Sci.* (2008) 33, p. 797
29. Nakata, N., Toda, T., Ishii, A., *Polym. Chem.* (2011) 2, p. 1597
30. Moritz, M. S., Baier, C., Zuideveld, M. A., Mecking, S., *Angew. Chem. Int. Ed.* (2014) 53, p. 9722
31. Rishina, L. A., Lalayan, S. S., Gagieva, S. C., Tuskaev, V. A., Shchegolikhin, A. N., Shashkin, D. P., Kissin, Y. V., *J. Res. Updates Polym. Sci.* (2014) 3, p. 216

4 Crystallinity Degree and Density of Polyethylene Resin

4.1 Crystallinity Degree

Most polyethylene resins are semi-crystalline plastics. Their crystallinity degree varies from 5 to 10% for VLDPE resins to ≥70% for some HDPE grades.

4.1.1 Measurement Methods

The crystallinity degree of polyethylene resins is measured with several physical methods including wide-angle X-ray spectroscopy (WAX), differential scanning calorimetry (DSC), infrared spectroscopy (IR), solid-state ^{13}C nuclear magnetic resonance (NMR), ultrasonic method, etc. [1–12].

Figure 4.1 shows the X-ray diffractogram of an ethylene/1-octene LLDPE resin [3]. The spectrum contains three prominent features: two narrow peaks of the crystalline phase of polyethylene, one at $2\theta = 21.20°$ (reflection from the <110> plane of the orthorhombic polyethylene cell) and another at 23.55° (reflection from the <200> plane); and a broad halo at $2\theta \sim 19°$ that is due to the amorphous phase. The crystallinity degree, X, is calculated from the areas under the two crystalline-phase peaks, S<110> and S<200>, and the area under the amorphous halo:

$$X, \% = 100\left(S<110> + K_1 \cdot S<200>\right) / \left(S<110> + K_2 \cdot S<200> + K_3 \cdot S<halo>\right) \quad (4.1)$$

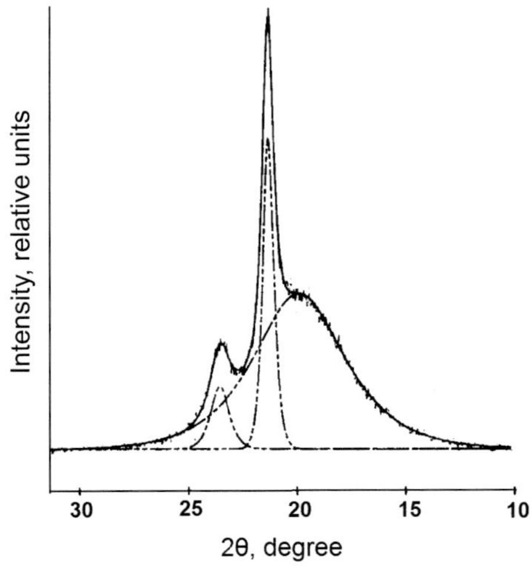

Figure 4.1
X-ray diffraction pattern of ethylene/1-octene LLDPE resin [3]

The values of parameters K_1, K_2, and K_3 are listed in Table 4.1.

Table 4.1 Parameters in Equation 4.1

K_1	K_2	K_3	Reference
1.43	1.43	0.69	[13, 14]
1.0	1.0	1.235	[2]
1.43	1.43	0.74	[12]
1.46	1.46	0.75	[1]

Figure 4.2 shows the DSC melting curve of an ethylene/1-hexene LLDPE resin. The heat of fusion, ΔH_f, is calculated as the area under the curve. The crystallinity degree is calculated as $X, \% = 100 \cdot (\Delta H_f/\Delta H_f^0)$ where ΔH_f^0 is the heat of fusion for perfectly crystalline polyethylene, $\Delta H_f^0 = 283$ to 302 J/g (67.6 to 72.0 cal/g) [3–6, 12, 15, 16]. The values of the crystallinity degree produced with the wide-angle X-ray and the DSC techniques correlate well in the range between ~10 and ~65%, the typical crystallinity range for MDPE, LLDPE, and VLDPE resins [3].

The ^{13}C-NMR spin relaxation measurement of a solid polyethylene resin gives magnetization decay curves. Each curve consists of three segments: the fast initial decay (T_1 ~0.3 sec) associated with the amorphous polymer component; the intermediate range with T_1 ~10 sec attributed to the interfacial material; and the range of slow decay due to the crystalline environment (T_1 ~100 to 300 sec) [3, 17–19]. Plotting these data in the coordinates "logarithm of the peak height as a function of time" gives the crystallinity degree as the intercept of the slowest decay range [3, 11, 18]. Automated techniques based on this method are widely used in industry for a rapid measurement of the crystallinity degree and density of polyethylene resins both in the pelletized and the granular form.

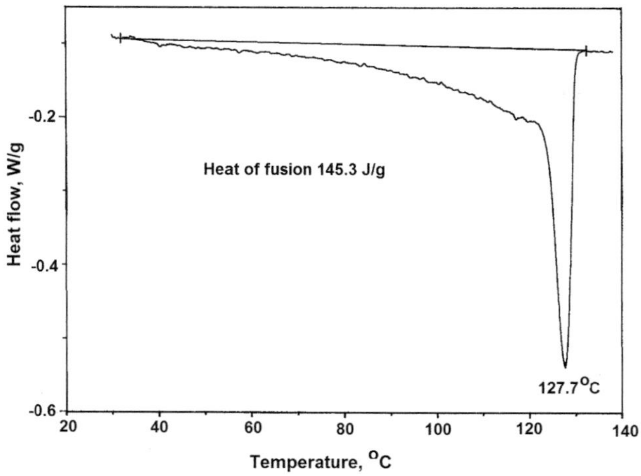

Figure 4.2 DSC melting curve of ethylene/1-hexene LLDPE resin, C_M^{copol} = 3.5 mol %

IR methods for the estimation of the crystallinity degree in press-molded annealed polyethylene film are based on the relative absorbance of the 1894 cm^{-1} band and different measures of the film thickness [2, 6, 12, 20]. If the film thickness is measured directly with a micrometer, the crystallinity degree is calculated as X, % = $100 \cdot K' \cdot A_{1894}/l$, where A_{1894} is the absorbance of the 1894 cm^{-1} band, l is the film thickness, in cm, and K' is the experimentally determined parameter, 16.4 [2] or 18.9 [12]. In other IR methods, the absorbance of the 1303 cm^{-1} band is used as the internal measure of the film thickness. In this case, the crystallinity degree is calculated as X, % = $100 \cdot (A_{1894}/A_{1303})/(A_{1894}/A_{1303} + K'')$ where K'' = 0.36 [2] or 0.30 [12].

4.1.2 Definition of Crystallinity Degree of LLDPE and VLDPE Resins Based on Copolymer Statistics

All commercially produced LLDPE and VLDPE resins are copolymers of ethylene with various α-olefins, 1-butene, 1-hexene, 1-octane, or 4-methyl-1-pentene (see Table 1.1). The molar content of an α-olefin in the copolymers, C_M^{copol}, is low, between 2.5 and 3.5 mol % in LLDPE resins and from 5 to 10 mol % in VLDPE resins.

As described in Section 1.7 of Chapter 1, any copolymer macromolecule containing ethylene units, E, and α-olefin units, M, can be regarded from the statistical viewpoint as a combination of monomer blocks of two types:

1. Ethylene blocks: $-M-(E)_n-M-$
2. α-Olefin blocks: $-E-(M)_m-E-$

with each block flanked by two monomer units of the opposite type.

Because the content of an α-olefin in LLDPE and VLDPE resins is relatively low, the average number n of monomer units in ethylene blocks M–$(E)_n$–M is quite high whereas the average number m of α-olefin units in blocks E–$(M)_m$–E is low; most of the α-olefin units are isolated in the chains as E–M–E sequences. As a result, the crystallinity degree and the melting point (Chapter 3) of all LLDPE and VLDPE resins are entirely determined by the presence of long ethylene sequences in the copolymer chains. The statistical parameter which defines the physical meaning of the crystallinity degree is the fraction of ethylene units in the sum of all long ethylene blocks starting with the block consisting of n units, that is, the fraction of ethylene units in the sum of all the blocks M–$(E)_n$–M, M–$(E)_{n+1}$–M, M–$(E)_{n+2}$–M, and so on.

$$\sum(E)_n = \left[\left(C_{\mathrm{M}}^{\mathrm{copol}}/100\right)\cdot(n-1)+1\right]\cdot\left\{\left[1-\left(C_{\mathrm{M}}^{\mathrm{copol}}/100\right)\right]\right\}^{n-1} \tag{4.2}$$

A comparison of the experimentally measured crystallinity degree of several compositionally uniform ethylene/1-hexene resins produced with a soluble metallocene catalyst and the computational results using Equation 4.2 is shown in Figure 4.3. The crystallinity degree of these copolymers was measured by the DSC method at a heating rate of 10 °C/min. Estimations of the n value were carried out under the assumption that the crystallization efficiency of polyethylene chains is approximately 0.6, a typical crystallinity level of rapidly crystallized HDPE resins determined from X-ray and DSC data. It follows from the plot that the size of the effective minimum length n of the crystallizable ethylene block is ~50 monomer units.

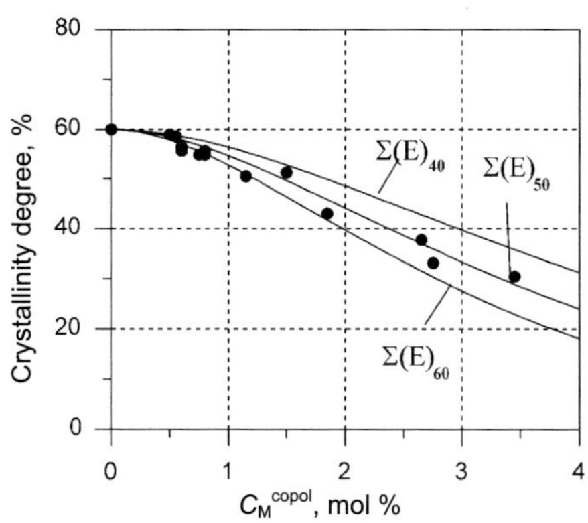

Figure 4.3
Comparison of experimentally measured crystallinity degree of compositionally uniform ethylene/1-hexene resins (points) and computational results produced with Equation 4.2

The packing efficiency of polyethylene chains depends on the crystallization rate of a molten resin. If the polymer samples are crystallized from the melt at a low cooling rate, 1 to 2 °C/min, and then melted at a rate of 2 to 3 °C/min (rather than 10 °C/min as in Figure 4.3), the packing efficiency increases to ~0.7 and, correspondingly, the effective minimum length n of the ethylene sequences decreases to ~40 ethylene units.

4.2 Density

Density is one of the primary characteristics of polyethylene resins. It is the basis of their classification (Chapter 1) and it is universally used as a singular parameter which instantly defines the application range of a given resin.

4.2.1 Measurement Methods

The density of polyethylene resins is measured by three methods, the flotation method in density-gradient columns, by the ultrasound method, and by the solid-state NMR method. The density measurement by the flotation method is carried out according to ASTM Methods D1505-10 and D2839-10. Binary liquid systems in density-gradient columns include the 2-propanol/water combination (suitable for the density measurement in the 0.79 to 1.00 g/cm^3 range) and 2-propanol/ethylene glycol combination (suitable for the 0.79 to 1.11 g/cm^3 range). The technique is very simple but it has a significant disadvantage: it requires a long sample annealing time for a precise measurement and the measurement itself takes several hours. At the present time, the ultrasonic method (ASTM D4883-08) and the solid-state NMR method (Section 4.1.1) are widely used for a rapid estimation of the polyethylene crystallinity degree and density.

4.2.2 Physical Meaning of Polyethylene Density

The physical definition of the density of ethylene/α-olefin copolymers is usually based on the simplest model of semi-crystalline materials. According to this model, any polyethylene resin consists of two phases, the crystalline and the amorphous. The model assumes that the values of the specific volumes for the crystalline and the amorphous fractions of the copolymers are additive:

$$1/d = 0.01 \cdot X/d_{cryst} + (1 - 0.01 \cdot X)/d_{amorph} \quad (4.3)$$

where X is the crystallinity degree, %, and d, d_{cryst}, and d_{amorph} are the density of the resin, the density of perfectly crystalline polyethylene, and the density of completely amorphous polyethylene, respectively. The d_{cryst} value is usually taken from the density data for linear alkanes (which produce perfect crystals), d_{cryst} = 1.0 g/cm^3. The d_{amorph} value cited in the literature varies in a narrow range from ~0.852 g/cm^3 (the density of highly branched isoalkanes) to 0.856 g/cm^3 [1, 2, 5, 12, 14–16, 21, 22].

One should take into account that articles manufactured from polyethylene nearly always contain two crystal modifications, the orthorhombic form (the dominant form) and the pseudo-monoclinic form (see Section 1.9). Theoretical density of these modifications is different; 1.0 g/cm^3 for orthorhombic polyethylene (the same as for crystalline linear paraffins) and 0.965 g/cm^3 for pseudo-monoclinic polyethylene. The pseudo-monoclinic form is stable only at temperatures below 50 °C and converts to the orthorhombic form at higher temperatures. To achieve reproducibility in the density measurement, a polyethylene sample should be annealed at a temperature from 80 to 100 °C, the procedure which converts the pseudo-monoclinic form into the orthorhombic form.

The model represented by Equation 4.3 has several shortcomings. First, it does not account for the fact that the presence of short-chain branches in ethylene/α-olefin copolymers leads to a small expansion of the **a** and the **b** axes of the orthorhombic cell and, hence, to a decrease of the theoretical d_{cryst} value compared to that for strictly linear polyethylene [3, 15, 21, 22]. Second, amorphous regions in the copolymers mostly contain linear sequences of ethylene units, $-CH_2-CH_2-$, with randomly positioned single or paired α-olefin units, $-CH_2-CHR-$, that is,

$-(CH_2-CH_2)_x-CH_2-CHR-(CH_2-CH_2)_y-$ and

$-(CH_2-CH_2)_x-CH_2-CHR-CH_2-CHR-(CH_2-CH_2)_y-$ sequences.

The density of such amorphous regions is significantly higher than that of highly branched isoalkanes.

To demonstrate difficulties associated with the density measurement of semi-crystalline polyethylene resins, Equation 4.3, after its transformation,

$$d = d_{cryst} \cdot d_{amorph} \cdot \left[d_{cryst} - 0.01 \cdot X\left(d_{cryst} - d_{amorph}\right)\right]^{-1} \tag{4.4}$$

was used to analyze the relationship between the crystallinity degree and density for several compositionally uniform ethylene/1-hexene copolymers produced with a single-site metallocene catalyst. Figure 4.4 presents the measurement results in the coordinates of Equation 4.4 (the crystallinity degree X was measured by DSC). The plot confirms that Equation 4.4 indeed represents the density-crystallinity dependence well. However, the data in the plot shows that the real density of the amorphous polyethylene phase is 0.877 g/cm^3 rather than 0.852 g/cm^3 and the density of the crystalline phase is 0.980 g/cm^3 rather than 1.00 g/cm^3. These differences emphasize significant problems associated with the use of the two-phase density model implicit in Equation 4.3 and Equation 4.4.

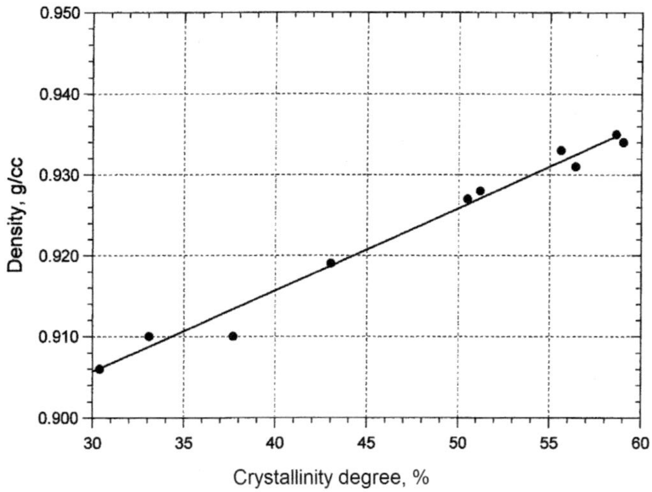

Figure 4.4 Dependence between density and crystallinity degree of compositionally uniform ethylene/1-hexene LLDPE and MDPE resins in coordinates of Equation 4.4

The density of any polyethylene resin (except for that of ethylene homopolymers) is primarily the function of the copolymer composition. Figure 4.5 gives one example of this dependence for the simplest case, compositionally uniform ethylene/1-hexene copolymers produced with a single-site metallocene catalyst. In theory, this dependence is equivalent to the dependence between the crystallinity degree and the copolymer composition shown in Figure 4.3. The slope of the line in Figure 4.5 primarily depends on the α-olefin content in a copolymer but it does not depend much on the nature of an α-olefin (except for propylene) [6, 23].

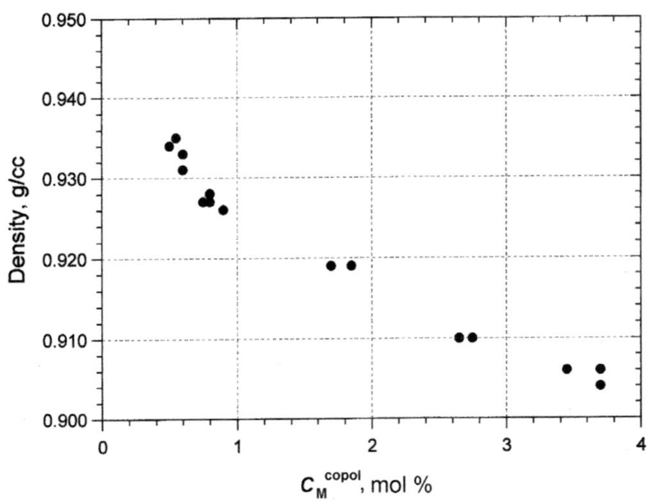

Figure 4.5 Density of compositionally uniform ethylene/1-hexene LLDPE and MDPE resins produced with metallocene catalyst as a function of 1-hexene content

Three other factors that affect the resin density are the thermal history of the sample, the molecular weight of the resin, and the degree of its compositional uniformity.

The role of the first factor is obvious: the faster the polyethylene melt crystallizes, the lower its crystallinity degree and, hence, its density. This uncertainty can be avoided by the controlled annealing of resin samples before the measurement.

The effect of the resin's molecular weight on its density is also quite significant. Common HDPE resins of high molecular weight (blow-molding resins with a high-load melt index I_{21} of ~1) have a maximum crystallinity degree of ~70%, and their maximum density is ~0.95 g/cm^3. Injection-molding HDPE resins of a lower molecular weight crystallize faster and their polymer chains are packed better in lamellae; their crystallinity degree is ~80% and density ~0.96 g/cm^3. In contrast, HDPE resins of ultrahigh molecular weight (over ~$2 \cdot 10^6$) have perfectly linear polymer chains but their crystallinity degree is low, ~30%, and their density is merely ~0.930 to 0.935 g/cm^3. In the case of LLDPE resins, the effect of the molecular weight (melt index) is usually accounted for empirically by adjusting the density of a given resin to the "apparent density" the resin would have if its I_2 melt index were 1.0 [24].

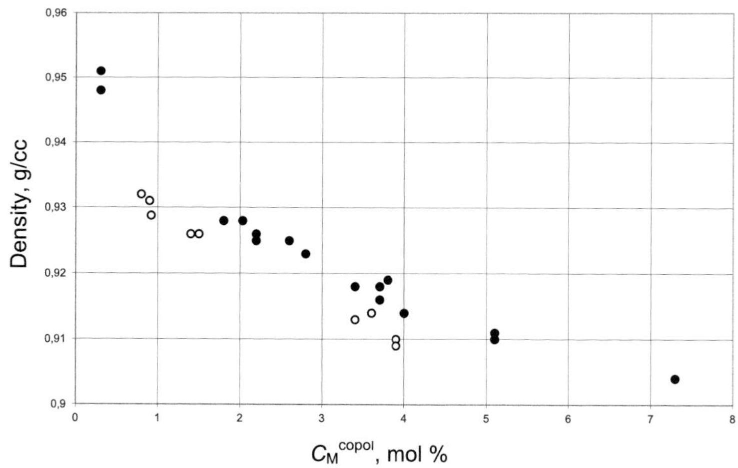

Figure 4.6 Density of ethylene/1-hexene LLDPE and MDPE resins produced with Ziegler-Natta catalyst as a function of 1-hexene content; [●] unfractionated resins; [○] narrow fractions

The last factor, the uniformity of compositional distribution in LLDPE resins, is also quite large and cannot be easily accounted for. To demonstrate its significance, Figure 4.6 shows the correlation between the density and the 1-hexene content for ethylene/1-hexene LLDPE and MDPE resins produced with a Ziegler-Natta catalyst and compares it with the data for narrow fractions of the same copolymers. The fractions were produced in a repeated extraction procedure and they are more uniform (although not perfectly) in terms of the compositional distribution. The density of the fractions is always noticeably lower than the density of unfractionated polymers of the same composition. The effect of compositional uniformity on the resin density can be clearly observed when the data in Figure 4.5 and Figure 4.6 are compared: the density of a compositionally uniform ethylene/α-olefin copolymer is always lower by ~0.01 to 0.012 g/cm^3 compared to the density of a compositionally nonuniform resin with the same average α-olefin content.

References

1. Aggarwal, S.L., Tilley, G.P., *J. Polym. Sci., Part B: Polym. Phys.* (1955) 158, p. 17
2. Hendus, H., Schnell, G., *KGK, Kautsch. Gummi Kunstst.* (1961) 51, p. 69
3. Clas, S.-D., Heyding, R.D., McFaddin, D.C., Russel, K.E., Scammell-Bullock, V.M., Kelusky, E.C., St-Syr, D., *J. Polym. Sci., Part B: Polym. Phys.* (1988) 26, p. 1271
4. Nowlin, T.E., Kissin, Y.V., Wagner, K.P., *J. Polym. Sci. Part A: Polym. Chem.* (1988) 26, p. 755
5. Ver Strate, G., Wilchinsky, Z.W., *J. Polym. Sci., Part A: Polym. Chem.* (1971) 9, p. 127
6. Beach, D.L., Kissin, Y.V., *J. Polym. Sci., Part A: Polym. Chem.* (1984) 22, p. 3027
7. Peacock, A., *Handbook of Polyethylene: Structures, Properties, and Applications* (2000) Google e-Book
8. Munaro, M., Akcelrud, L., *J. Polym. Res.* (2011) 15, p. 83
9. Lin, W., Cossar, M., Dang, V., *J. Polym. Test.* (2007) 26, p. 814
10. Standard Test Method for Density of Polyethylene by the Ultrasound Technique (ASTM), D4883-08, *http://www.astm.org/Standards/D4883.htm*
11. Kitamaru, R., Horii, F., Murayama, K., *Macromolecules* (1986) 19, p. 636
12. Mirabella, F. M., A. Bafna, A., *J. Polym. Sci., Part B: Polym. Phys.* (2012) 40, p. 1637
13. Nichols, J.B., *J. Appl. Phys.* (1954) 25, p. 340
14. Munos-Escalona, A., Parada, A., *Polymer* (1979) 20, p. 859
15. Bodily, D., Wunderlich, B., *J. Polym. Sci., Part A: Polym. Chem.* (1966) 4, p. 25
16. Runt, J., Harrison, I.R., Dobson, S., *J. Macromol. Sci., Part B. Phys.* (1980) B17, p. 99
17. Axelson, D.E., Mandelkern, L., Popli, R., Mathieu, P., *J. Polym. Sci., Part B: Polym. Phys.* (1983) 21, p. 2319
18. Axelson, D.E., Russell, K.E., *Prog. Polym. Sci.* (1985) 11, p. 221
19. Dujourdy, I., Bazile, J.R., Cohen-Addad, J.P., *Polym. Int.* (1999) 38, p. 558
20. Okada, T., Mandelkern, L., *J. Polym. Sci., Part A: Polym. Chem.* (1967) 5, p. 239
21. Vonk, C.G., In *Integration of Fundamental Polymer Science and Technology*. Lemstra, P.J., Kleintjens, L.A. (Eds.) (1988) Elsevier Applied Science, London
22. Vonk, C.G., Reynaers, H., *Polym. Commun.* (1990) 31, p. 190
23. Dreval, V.Y., Artamonova, S.D., Bobrov, B.N., Kleiner, V.I., Barancheyeva V.V., Litvinov, I.A., Krechekyan, A.S., *Polym. Sci.* (1991) 11, p. 2461
24. Elston, C.T., *U.S. Patent* 3 645 882 (1972)

5 End-Use Mechanical Properties of Polyethylene Film

End-use tests of blown and cast film manufactured from high density polyethylene (HDPE), low density polyethylene (LDPE), and linear low density polyethylene (LLDPE) usually include two mechanical tests described in this chapter, the dart impact test and the tear test [1–4], as well as the tests on the amount of soluble material and optical properties (haze), which are described in Chapter 7. The results of the two mechanical tests are used in industry to grade the resins, determine their range of applications and price, and compare polymerization catalysts and processes.

In the case of LLDPE resins (ethylene/α-olefin copolymers containing 2.5 to 3.5 mol % of α-olefin), both the dart impact strength and the tear strength of the film are affected by numerous factors, such as the type of the α-olefin, its content in the copolymer, average molecular weight, and compositional uniformity [5–7]. Although general information on mechanical properties of polyethylene resins is readily available in commercial publications, various subtle effects of the resin's structure on the end-use film properties – the effects that are often paramount for successful applications of a given resin – are relatively poorly understood. A better understanding is hindered by the physical complexity of these two seemingly straightforward mechanical tests and also by the interaction of various structural factors of the resins that affect the test results.

5.1 Mechanical Properties of Polyethylene Resins

Section 1.9 and Section 1.10 in Chapter 1 describe principal morphological features and basic mechanical properties of semi-crystalline polyethylene resins. Most of these resins are materials with expressed ductile properties. Morphological and mechanical changes in the resins subjected to stretching are represented by the stress/strain curve shown in Figure 1.18. The three defining points on the curve are the yield point, the end-of-necking point, and the breaking point. These points are characterized by six parameters:

1. The yield stress, σ_y, and the yield strain, ε_y;
2. The necking stress, σ_n, and the end-of-necking strain, ε_n;
3. The breaking (tensile) stress, σ_{br}, and the breaking strain, ε_{br}.

The ε values in Figure 1.18 are the length ratios in the strained and the original sample: ε = elongation + 1; the minimum ε value for a polymer sample before stretching is 1.

Mechanical changes shown in Figure 1.18 are accompanied by irreversible structural changes in the polymers; they are schematically shown in Figure 1.19. After the yield point of a polymer sample is passed, an area consisting of a highly oriented material (the neck) develops in the strained sample. As the stretching proceeds further, the two principal morphological features of a semi-crystalline resin, the spherulites and the microfibrils (see Figure 1.16 and Figure 1.17), are gradually disassembled until all the material in the tested sample becomes highly oriented. This transformation occurs at a nearly constant necking stress σ_n. The last stage of the sample stretching is called strain hardening; it involves a simultaneous further increase in the strain and the stress in the oriented sample. At some point during this stage the material finally breaks. The length of the strain-hardening range, $\varepsilon_{br} - \varepsilon_n$, depends on the grade of the resin. HDPE resins break soon after the onset of strain hardening; their σ_y and σ_{br} values are relatively close. On the other hand, LLDPE resins have a relatively large strain-hardening range and their σ_{br} value is always much higher than the σ_y and the σ_n values.

Two more mechanical parameters are used to characterize polyethylene resins. The first parameter, the Young's modulus (M_{Young}), characterizes stiffness of a resin sample, the slope of the stress/strain curve at very low strain. By definition, $M_{\text{Young}} = d(\sigma)/d(\varepsilon - 1)$ at low ε. In practice, two similar parameters, the secant modules, are often used, the stress at 1% elongation or at 2% elongation. The second mechanical parameter, the strain-hardening modulus ($M_{\text{str-hard}}$), characterizes stiffness of a fully stretched and oriented resin sample: $M_{\text{str-hard}} = (\sigma_{br} - \sigma_n)/(\varepsilon_{br} - \varepsilon_n)$.

5.1.1 Effect of Stretching Speed on Mechanical Properties

Both end-use mechanical tests of polyethylene film, the dart impact test and the tear test, are carried out at a high deformation speed (Section 5.2 and Section 5.3). Relaxation phenomena in polyethylene are relatively slow, and most mechanical parameters of the resins depend on the deformation speed [8, 9]. These dependencies were measured experimentally in a broad range of deformation speeds V from 0.5 to 150 cm/min (0.2 to ~60 in/min) for several ethylene/1-hexene LLDPE resins, both compositionally uniform resins produced with metallocene catalysts and compositionally nonuniform resins produced with Ziegler-Natta catalysts. In most cases, the dependencies between the tested parameters, σ or ε, and the deformation speed can be represented by simple empirical relationships using as a standard a particular σ or ε value at the standard deformation speed V_{stand} of 50.8 cm/min (20 in/min):

$$\sigma_y(V) = \sigma_y^{\text{stand}} + k(\sigma_y)\cdot\log(V/V_{\text{stand}}),\quad \text{the slope } k(\sigma_y) \approx 0.14 \tag{5.1}$$

$$\sigma_n(V) = \sigma_n^{\text{stand}} + k(\sigma_n)\cdot\log(V/V_{\text{stand}}),\quad \text{the slope } k(\sigma_n) \approx 0.11 \tag{5.2}$$

$$\varepsilon_n(V) - \varepsilon_y(V) = \varepsilon_n^{\text{stand}} - \varepsilon_y^{\text{stand}} + k(\varepsilon_n - \varepsilon_y)\cdot\log(V/V_{\text{stand}}) \tag{5.3}$$

The slope $k(\varepsilon_n - \varepsilon_y)$ in the last expression varies from 0.5 to 0.6 depending on the LLDPE grade. Three other parameters of the stress/strain curve, ε_y, ε_{br}, and σ_{br}, practically do not depend on the deformation speed.

5.1.2 Orientation in Polyethylene Film

Figure 5.1 shows the orientation pattern, orientation of the c axis in crystallites (the direction of molecular chains) in two types of articles manufactured from polyethylene, a blow-molded item and film.

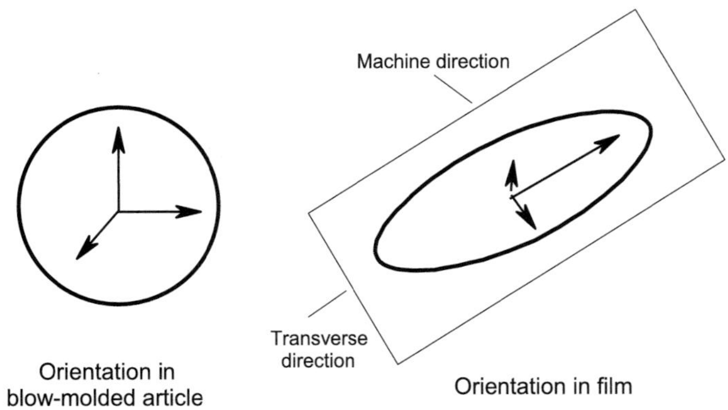

Figure 5.1 Orientation distribution in blow-molded article and in film

Thick-walled, blow-molded articles are practically isotropic in terms of chain orientation. On the other hand, polymer chains in polyethylene film are preferably oriented in the machine direction of film manufacture. The level of orientation depends on the resin type; it is the highest in HDPE film but significantly lower in film made of LLDPE resins. Several examples of the orientation level in film are given in Table 5.1.

Table 5.1 Chain Orientation in Crystalline Phase of Polyethylene Film; IR Measurements [14]

Resin type and density	Chain orientation in crystallites		
	Machine direction	Transverse direction	Thickness direction
HDPE, 0.96 g/cm^3	0.59 to 0.63	0.39 to 0.35	~0.02
LLDPE, ethylene/1-butene, 0.923 g/cm^3	0.51	0.44	~0.05
LLDPE, ethylene/1-hexene, 0.918 g/cm^3	0.44	0.35	~0.20

The preferred chain orientation in the film machine direction explains large differences in the film's mechanical properties, which play an especially important role in the measurement of the tear strength (Section 5.3). As an example, Table 5.2 gives experimental data on ε_y, ε_n, and ε_{br} values of ethylene/1-butene LLDPE film slowly stretched in the machine and the transverse directions, and Figure 5.2 shows experimental stress/strain curves of blown ethylene/1-hexene LLDPE film recorded at a stretching speed of 2.1 cm/sec (0.83 in/sec).

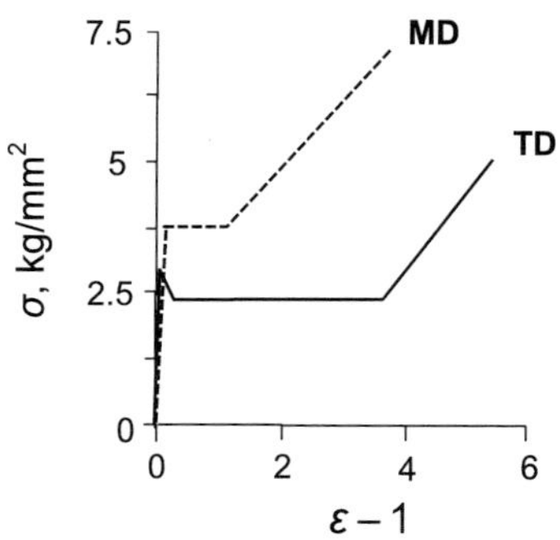

Figure 5.2
Stress/strain curves of blown ethylene/1-hexene LLDPE film (density 0.918 g/cm^3) measured in machine direction (MD) and transverse direction (TD); stretching speed 2.1 cm/sec

Mechanical properties of the film in the two directions are obviously different: σ_{br}(machine) > σ_{br}(transverse) and ε_n(machine) < ε_n(transverse), reflecting differences in the chain orientation. The strain-hardening modules of the film, $M_{\text{str-hard}} = (\sigma_{br} - \sigma_n)/(\varepsilon_{br} - \varepsilon_n)$ are also noticeably different in the two directions: $M_{\text{str-hard}}$(machine) ~ 1.3 kg/mm^2 and $M_{\text{str-hard}}$(transverse) ~ 1.8 kg/mm^2. Several examples of such measurements for different grades of LLDPE film showed the same trends [1–7]. The highest degree of orientation in the machine-direction was observed in film manufactured from HDPE resins of ultrahigh molecular weight (M_w over $1.5 \cdot 10^6$). These resins can be processed at a high pressure into film that is very strongly stretched and nearly perfectly oriented in the extrusion direction, both in the crystalline and the amorphous phases [10, 11]. The data for molecular orientation in HDPE tubing and pipes are not available, but related data for polypropylene pipes also show a noticeable (although relatively small) degree of preferred molecular orientation in the extrusion direction [12].

Table 5.2 Elongation Ratios for LLDPE Film (Ethylene/1-Butene Copolymer, Density 0.918 g/cm^3) at Low Stretching Speed

Stretching direction	ε_y [a]	ε_n [a]	ε_{br}
Machine	1.2 to 1.3	3.0 to 3.2	4.0 to 4.2
Transverse	1.3 to 1.4	5.5 to 6.0	6.5 to 7.0

[a] Relaxation to original length takes several seconds

5.2 Dart Impact Strength of LLDPE Film

5.2.1 Description of Dart Impact Test

The procedure of the dart impact (dart drop) testing according to the American Society for Testing and Materials (ASTM) D1709-04 and the International Organization for Standardization (ISO) 7765 is simple [13]. It is shown in Figure 5.3. A dart with a rounded tip with a curvature radius of ~16 mm (~0.63 in) and a thin long stem is loaded with several cylindrical metal weights and dropped from a height of 660 mm (26 in) perpendicularly to the surface of a round piece of polymer film [diameter 127 mm (5 in)] fastened between two metal rings. The weight of the dart can be varied over a wide range by changing the number of weights on the stem. The film is usually tested about 20 times at different dart weights until its partial or complete rupture occurs. A special protocol for varying the dart weight and calculating the ratio of failure to survival incidents gives the average dart weight needed to break the film. This averaged dart weight value, in grams, is called the dart impact strength [13]. Test results for different grades of polyethylene film vary over a wide range, from ~50 to 70 to over 1,000 g. The second ASTM test, ASTM D4272, uses a single heavy dart and measures the loss of the dart's kinetic energy by comparing the time of the dart's fall in the test and the time of its free fall in the absence of film.

Mechanical processes occurring during a dart impact test can be clearly seen when the tests are performed on film manufactured from a high-quality LLDPE resin. Figure 5.3 shows schematics of three stages of the test. The dart approaches the film at a speed V_0 of ~3.6 m/sec (~142 in/sec). The central part of the film, 22 to 25 mm in diameter, clings to the semispherical surface of the dart head and remains undamaged during the test. At the first stage of the test, the dart bends the film until the stress at the circumference of the central (undamaged) film area exceeds the yield stress of the resin. After that, the area around the undamaged area begins to stretch (Stage 2 in Figure 5.3). The work of stretching decreases the kinetic energy of the dart, and its speed rapidly decreases (see the quantitative estimation below). If the kinetic energy of the dart is insufficient to break the film, the dart stops inside the cavity of the stretched film (Stage 2). In the case of film manufactured from a high-quality resin, the depth of the cavity can reach 35 to 40 mm (1.4 to 1.6 in). If the kinetic energy of the dart exceeds the work needed to break the film, the dart makes a round hole in the film and falls through, carrying with it the central undamaged part of the film clinging to its surface (Stage 3). The break line is usually located very close to the circumference of the undamaged film area.

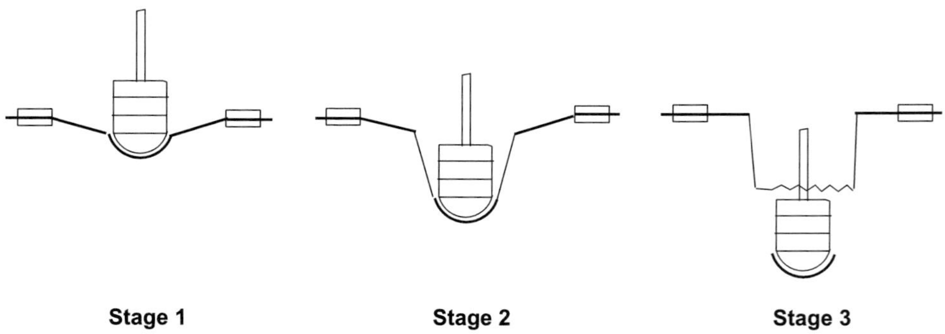

Figure 5.3 Three stages of dart impact test of semi-crystalline film

Deceleration of dart during the test: The falling dart decelerates rapidly during the test; its starting speed, V_0, is high, ~3.6 m/sec (~142 in/sec), but it is reduced to the final speed $V_L = 0$ in a fraction of a second. The general expression for the dart speed as a function of the length l of the film subjected to stretching is [9]:

$$V_t = V_0 \cdot \{1 - 2 \cdot (l/d) \cdot (1 + 2 \cdot (l/d)) / [(\sigma_{br}/\sigma_n) \cdot (\sigma_{br}/\sigma_n - 1)]\}^{0.5} \quad (5.4)$$

Figure 5.4 gives several examples of the dart speed ratio V_l/V_0 as a function of the length l of the film subjected to irreversible stretching. The calculations were carried out for five σ_{br}/σ_n ratios typical for LLDPE resins of various types. The figure shows that the final stage of the dart impact test, immediately before the film rupture, proceeds at a low speed.

The size of the undamaged film area (diameter d), which clings to the tip of the dart and is torn off when the film fails (Figure 5.3), was estimated in dart impact experiments with several types of LLDPE film. The tests were carried out with widely varying dart weights, from very small, barely sufficient to initiate the film stretching, to very large, exceeding the dart impact strength of the films. The tests showed that the undamaged area has an elliptical shape with the axis ratio of 1.1 to 1.2.

The reason for the elliptical rather that the circular shape is the orientation of the material during the film manufacture (Section 5.1.2); the mechanical parameters in the machine direction and the transverse direction of the film are different. The average diameter of the undamaged area is practically independent of the dart weight and of the degree of damage the dart does to the film, see Table 5.3. Similar results were produced with other types of 25 µm- and 38 µm-thick films prepared from ethylene/1-butene, ethylene/1-hexene, ethylene/1-octene, and ethylene/4-methyl-1-pentene resins; the *d* values range from 22.5 to 24.1 mm.

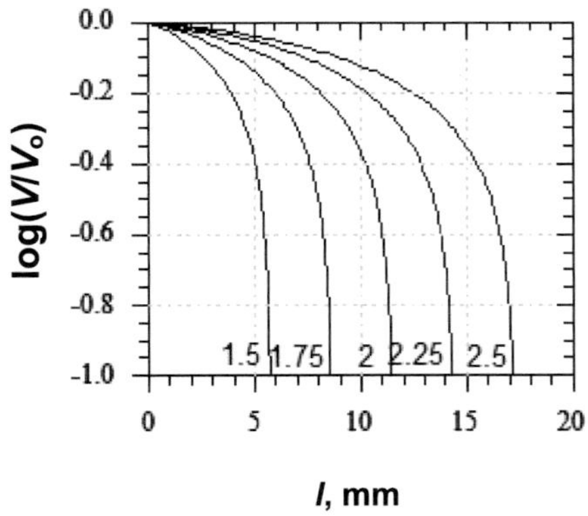

Figure 5.4 Decrease of dart speed as a function of deformed polymer zone for five different σ_{br}/σ_n ratio

Table 5.3 Effect of Dart Test Parameters on Diameter of Undamaged Zone in LLDPE Film (Density 0.917 g/cm^3; I_2 = 1.0 g/10 min)

Dart weight, g:	80	120	240	280
Film condition:	not broken	failed	failed	failed
***d*,[a] mm:**	23/25	23/25	23/26	22/25

[a] Minor and major axis of the ellipse

Stretching of the film during the dart impact test is accompanied by strong orientation in the polymer sleeve created by the falling dart (Figure 5.3, Stages 2 and 3). Analysis of polarized infrared spectra of the stretched area after the test showed strong orientation of polyethylene molecules in the direction of the sleeve. The degree of chain orientation in the test direction is comparable to that achieved by stretching the same film at a high speed to the end-of-necking point [14].

These details of the dart impact test show that the conditions of the test differ from the conditions of the standard tensile test in two respects:

1. The stretching speed during the dart impact test varies; it is high at the beginning of the test and sharply decreases at the end of the test. If the dart weight is slightly higher than the dart impact strength of the film, stretching of the film at the final moments of the test, immediately before the film failure, proceeds at a

relatively low speed. If the dart weight is insufficient to break the film, the final stretching speed is zero.

2. Due to the geometrical design of the test, the cross section of the stretching area continuously increases as the film stretching progresses.

5.2.2 Model of Dart Impact Test

The idealized geometry of a film sample undergoing the dart impact test is shown in Figure 5.5(a). Three parameters describe the sample: its thickness t (mm), the diameter of the test area D (mm), and the diameter of the central undamaged area of the film clinging to the tip of the dart, d (mm). In the beginning of the test, the stretching starts at the circumference of the cling area (Stage 1 in Figure 5.3). The initial cross section of the stretching zone at this point is the smallest, $S_{min} = \pi \cdot d \cdot t$. As the film stretching progresses, the cross section of the stretched zone increases in a linear manner starting from S_{min}. When the stretching proceeds to a distance l (assigned to the original film before the test), the cross section of the film passing through the yielding point is $S_l = \pi \cdot (d + 2 \cdot l) \cdot t$. The maximum possible stretching cross section, which is never achieved in real dart impact tests, is $S_{max} = \pi \cdot D \cdot t$.

The model of the dart impact test is based on two definitions:

1. The dart impact strength of a film sample, m_{dart} (gram), is defined as the maximum weight of the dart the film can stop without breaking, or as the minimum weight of the dart needed to break through the film.
2. Tensile properties of LLDPE resins can be adequately described by the idealized standard stress/strain curve shown in Figure 5.5(b).

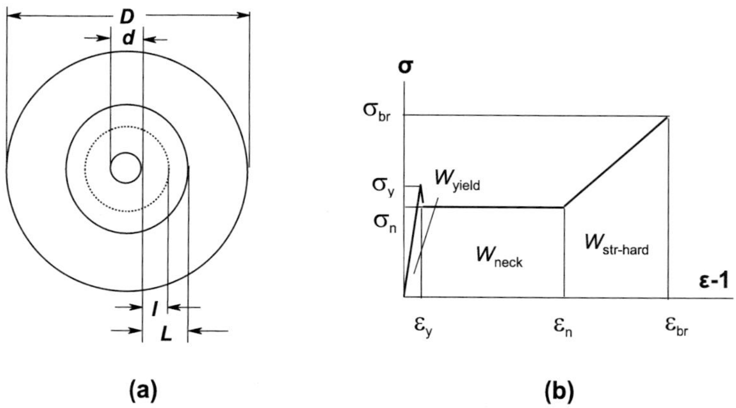

Figure 5.5 (a) Geometry of dart impact testing; (b) idealized stress/strain curve of polymer sample

The film breaks when the stress in the smallest cross section, S_{min}, reaches the breaking stress of the material, σ_{br}. The breaking force F_{br} at this moment is:

$$F_{br} = \sigma_{br} \cdot S_{min} = \sigma_{br} \cdot \pi \cdot d \cdot t \tag{5.5}$$

The same force F_{br} is applied to all cross sections of the sample. At this point, the necking zone propagates along the film radius to the maximum distance L (assigned to the original film, $L < D$); it is the distance X in the stretched film after stress relaxation, $X/L = \varepsilon_n$). The relationship between L and two mechanical parameters of the resin, σ_y and σ_{br}, is defined as:

$$F_{br} = \sigma_y \cdot \pi(d + 2 \cdot L) \cdot t = \sigma_{br} \cdot \pi \cdot d \cdot t \tag{5.6}$$

Equation 5.6 signifies that at the moment when the film fails, the yield stress σ_y is achieved at the distance L from the undamaged zone, at the point where the cross-section area is $S_L = \pi \cdot (d + 2L) \cdot t$. Solution of Equation 5.6 gives the following expressions for L and X:

$$L = (\sigma_{br} - \sigma_y) \cdot d/(2 \cdot \sigma_y) \text{ and } X = \varepsilon_n \cdot (\sigma_{br} - \sigma_y) \cdot d/(2 \cdot \sigma_y) \tag{5.7}$$

According to the definition of the dart impact strength, the weight of the dart sufficient for the film rupture, m_{dart}, is determined by the expression that equates the kinetic energy of the dart immediately before the impact and the total work of film stretching and breaking, W_{total}:

$$m_{dart} \cdot V^2/2 = W_{total} \tag{5.8}$$

where V = ~ 3.6 m/sec (~ 142 in/sec).

The total work required to stretch the film to the breaking point can be separated into four parts:

$$W_{total} = W_{yield} + W_{elast} + W_{neck} + W_{str\text{-}hard} \tag{5.9}$$

where W_{yield} is the work of elastic stretching to the yield point in all cross sections from S_{min} to S_L (Figure 5.5(b)); W_{elast} is the work of elastic stretching in all cross sections from S_L to S_{max}; W_{neck} is the work of necking (i.e., stretching/orientation of the film material from $l = 0$ to L); and $W_{str\text{-}hard}$ is the work of strain hardening, the additional stretching of the oriented part of the sample to the point when the stress in the smallest cross section reaches σ_{br}.

The second dart impact test, ASTM D4272 [13], measures the loss of the kinetic energy, ΔE, of the dart breaking the film by comparing the time of the dart fall in the test to the time of its free fall in the absence of film. When all the σ values are presented in kg/mm^2, all cross sections in mm^2, and all distances in mm, the ΔE value for a given film can be estimated as $\Delta E = 9.81 \cdot 10^{-3} \cdot W_{total}$ (J).

Figure 5.5(b) shows three components of the total work required to break the film for a polymer sample with a constant cross section. Analysis of mechanical testing in the case of a film sample with a varying cross section gives the following expressions for each term in Equation 5.9 [9]:

$$W_{\text{yield}} = 0.125 \cdot \pi \cdot d^2 \cdot t \cdot \sigma_y \cdot (\varepsilon_y - 1) \cdot \left[(\sigma_{br}/\sigma_y)^2 - 1\right] \tag{5.10}$$

$$W_{\text{elast}} = 0.25 \cdot \pi \cdot t \cdot \sigma_y \cdot (\varepsilon_y - 1) \cdot (d + 2L)^2 \cdot \ln[D/(d + 2L)] \tag{5.11}$$

$$W_{\text{neck}} = 0.25 \cdot \pi \cdot \sigma_n \cdot (\varepsilon_n - \varepsilon_y) \cdot d^2 \cdot t \cdot \left[\left(\sigma_{br} / \sigma_y\right)^2 - 1\right] \tag{5.12}$$

$$W_{\text{str-hard}} = (1/6) \cdot \pi \cdot t \cdot (\varepsilon_{br} - \varepsilon_n) \cdot L \cdot [d \cdot (2 \cdot \sigma_{br} + \sigma_n) + L \cdot (\sigma_{br} + \sigma_n)] \tag{5.13}$$

Obviously, the work of the dart impact is a complex function of all six parameters that characterize the stress/strain curve of a resin, the stresses and the strains at the yield, the end-of-necking, and the breaking points.

The simple model of the dart impact test represented by Equation 5.9 to Equation 5.13 has one obvious shortcoming. It does not take into account the large decrease in the stretching speed of the film from the first moment, when the dart speed V is ~ 3.6 m/sec (~ 142 in/sec), to the end-point of the test, when the dart speed $V \approx 0$, and the effect of the dart speed change on mechanical properties of the film. The only term in Equation 5.9 that requires a significant correction to account for this effect is the W_{neck} term in Equation 5.12 [9].

Calculations of the dart impact strength show that different terms in Equation 5.9 contribute to the impact strength of the film to a different degree. As an example, the following mechanical parameters were chosen to characterize a typical ethylene/1-hexene LLDPE resin:

$\sigma_y \sim 1.3$ kg/mm^2, $\varepsilon_y \sim 1.2$, $\sigma_n \sim 1.2$ kg/mm^2, $\varepsilon_n \sim 6.0$, $\sigma_{br} \sim 2.3$ kg/mm^2, $\varepsilon_{br} \sim 7.5$.

The estimation of the dart impact strength for a 25-µm (1-mil) film prepared from a resin with these properties gives $m_{\text{dart}} \sim 255$ g. This estimation compares well to the experimental data, which shows the dart impact strength of film prepared from such resins from 150 to 200 g. The energy of the falling dart is spent in the following way: 1.5% for stretching the film to the yield point; 5.5% for elastic deformation of the nonoriented part of the film; 76% for necking/orientation of the film; and 17% for overcoming the resin's strain hardening.

Table 5.4 shows the estimated effects of the sample diameter, D, and the diameter of the undamaged area, d, on the dart impact strength. A change in D over a wide range has virtually no effect on the calculated test results, provided that $D \gg d$. On the other hand, the size of the undamaged area, d, strongly affects the film strength value. However, experimental results in Table 5.3 show that this parameter is mostly independent of the resin's properties. Also, according to the model, the dart impact strength is proportional to the film thickness (Equation 5.10 to Equation 5.13).

Table 5.4 Effects of Test Parameters on Calculated Dart Impact Strength of 25-μm (1-mil) Film[a]

Test parameters		Dart impact strength
D, mm	*d*, mm	m_{dart}, g
100	23	241
125	23	244
150	23	246
127	20	186
127	23	244
127	26	410

[a] Resin parameters: σ_y = 1.1 kg/mm², σ_n = 1.0 kg/mm², σ_{br} = 2.0 kg/mm², ε_y = 1.2, ε_n = 6, ε_{br} = 8

The model of the dart impact test has several deficiencies. It describes a perfect film break taking place along the circumference of the cling zone (Figure 5.3). In reality, such a break is observed only for compositionally uniform resins, which produce weakly oriented film. Most other types of blown LLDPE film become partially oriented in the course of manufacture. When a dart impacts such a film, it often produces one or two cracks stretched in the machine direction of the film tangentially to the cling zone. The formation of the cracks qualifies such tests as a failure, although no complete breaking of the film really occurs. The degree of orientation in the film (and the probability of film cracking) increases with decreasing α-olefin content in the copolymers, increasing the resin crystallinity degree (density), and increasing the tendency to orientation during the film manufacture. This phenomenon explains the discrepancy between high dart impact strength predictions for medium density polyethylene film and its relatively low experimentally measured "dart impact strength," which in reality should be called "dart-cracking strength."

5.2.3 Effect of Mechanical Properties of LLDPE Resins on Dart Impact Strength

It follows from Equation 5.10 to Equation 5.13 that all the six parameters characterizing the stress/strain curve of an LLDPE resin, σ_y, ε_y, σ_n, ε_n, σ_{br} and ε_{br}, affect the film dart impact strength in a complex way due to their contributions to different terms in Equation 5.9. Computational results for 25-μm (1-mil) film in Table 5.5 demonstrate the effects of LLDPE mechanical properties on the calculated dart impact strength. The following trends can be observed:

1. Existence of a pronounced yield threshold, $\sigma_y/\sigma_n > 1$, which is typical for many LLDPE resins (provided that the values of σ_n, σ_{br}, ε_y, ε_n, and ε_{br} remain the same), is detrimental to the dart impact strength, although this effect is not large in the typical range of the σ_y/σ_n ratio, 1.05 to 1.10.
2. The value of the Young's modulus, M_Y, has virtually no effect on the dart impact strength of the film.
3. An increase of resistance to orientation (stiffness, a proportional increase of σ_y and σ_n) produces a strong decrease in the dart impact strength. This trend explains why soft and easily stretchable LLDPE resins with a uniform compositional distribution (for example, the resins produced with metallocene catalysts) exhibit a superior dart impact strength.
4. An increase of the orientation potential of a resin (an increase in ε_n) does not affect the dart impact strength.
5. An increase of the resin tensile strength, σ_{br}, greatly improves the dart impact strength of the film by allowing the orientation zone in the film to propagate further, to higher l. An increase in the strain-hardening properties of a resin (a parallel increase of σ_{br} and ε_{br}) produces a similar effect.

Table 5.5 Effects of LLDPE Mechanical Properties on Dart Impact Strength

Effect of yield threshold σ_y/σ_n at σ_n = 1.0 kg/mm², σ_{br} = 2.5 kg/mm², ε_y = 1.2, ε_n = 5, ε_{br} = 7				
σ_y/σ_n:	1.02	1.05	1.10	1.15
m_{dart}, g:	330	318	296	273
Effect of Young's modulus, $M_Y = \sigma_y/(\varepsilon_y - 1)$				
$\sigma_y/(\varepsilon_y - 1)$:	7.3	5.5	4.5	3.4
m_{dart}, g:	293	296	299	304
Effect of resistance to orientation (proportional increase of σ_y and σ_n at σ_y/σ_n = 1.1)				
σ_n, kg/mm²:	0.8	1.0	1.2	1.4
m_{dart}, g:	500	296	217	152
Effect of orientation (necking) potential				
ε_n:	4.0	6.0	8.0	
m_{dart}, g:	298	294	289	
Effect of increase of tensile strength				
σ_{br}, kg/mm²:	2.0	2.5	3.0	
m_{dart}, g:	203	372	764	

5.2.4 Comparison of Film Made from Ethylene/1-Butene and Ethylene/1-Hexene Copolymers

LLDPE film manufactured from ethylene/1-butene copolymers is inferior in dart impact strength compared to film manufactured from ethylene/1-hexene copolymers of the same composition and crystallinity degree (density). The average dart impact strength of 25-µm blown film made from an ethylene/1-butene resin is 80 to 120 g, whereas this parameter for blown film from a common ethylene/1-hexene resin is 150 to 200 g and it can reach 400 to 500 g for film from high-strength ethylene/1-hexene copolymers prepared with special types of Ziegler-Natta catalysts [15].

Table 5.6 lists mechanical properties of several resins of both types. The two ethylene/1-butene resins and the first of the ethylene/1-hexene resin were produced with the same catalyst; they are similar in density, degree of crystallinity, and molecular weight. The resins have similar yield and necking parameters, but differ noticeably in tensile properties: the ethylene/1-butene copolymers have a ~30% lower breaking stress. The results in Table 5.6 can be interpreted in the following way. The work required to start a neck, W_{yield}, and the work required to stretch elastically the undeformed part of the film, W_{elast}, are practically the same for both types of film because their yield parameters, σ_y and ε_y, are similar. The film manufactured from an ethylene/1-butene resin breaks easier in the dart impact test primarily because the breaking stress σ_{br} of this resin is lower. As a result, the work of necking/orientation, W_{neck}, for the film from the ethylene/1-butene resin is two times lower. The work required to overcome strain hardening, $W_{str\text{-}hard}$, for the film from an ethylene/1-butene copolymer is also lower by a factor of 1.5 to 2.

Table 5.6 Mechanical Properties and Estimated Dart Impact Strength of 25-µm (1-mil) Film from Compositionally Nonuniform Ethylene/α-Olefin Resins (Density 0.918 g/cm^3)

Resin[a]	Mechanical properties[b]							Estimated dart impact parameters[b]				
	σ_y	ε_y	σ_n	ε_n	σ_{br}	ε_{br}	m_{exp}	m_{calc}	W_{yield}	W_{elast}	W_{neck}	$W_{str\text{-}hard}$
E/B	1.01	1.27	0.94	6.4	1.53	9.9	~100	160	2	8	67	29
E/B	1.03	1.36	0.98	5.1	1.56	8.9	~120	105	2	14	39	14
E/H	1.20	1.45	1.04	5.7	2.39	8.7	~200	320	7	23	136	43
E/H	1.24	1.40	1.05	5.3	2.94	8.7	~580	640	12	25	229	157

[a] E/B – ethylene/1-butene resins, E/H – ethylene/1-hexene resins
[b] All σ values are in kg/mm^2, all W values are in kg · mm, both m_{dart} values are in g

The last two rows in Table 5.6 compare experimental data and computational results for film made from two different ethylene/1-hexene copolymers, the first one a resin produced with a common supported Ziegler-Natta catalyst and another one a "super-strength" resin produced with a modified Ziegler-Natta catalyst [15]. The two resins are identical in most of their tensile parameters. The superior dart impact strength of film made from for the "super-strength" resin is mostly determined by a ~20% higher breaking stress σ_{br} value, and, respectively, by a higher contribution of W_{neck} and $W_{str\text{-}hard}$ terms in Equation 5.9 to the W_{total} value.

5.2.5 Effect of Copolymer Composition

The dart impact strength of film from all ethylene/α-olefin resins of a similar average molecular weight significantly increases as the α-olefin content in them increases and their density and the crystallinity degree decrease [1, 2, 5, 6]. The data in Table 5.7 provide an explanation for this trend. As the 1-butene content in an ethylene/1-butene copolymer increases from 3.5 to 4.4 mol %, all the stresses (σ values) of the material decrease while all the strains (ε values) remain essentially the same. Paradoxically, the film manufactured from the mechanically weaker material has a higher dart impact strength (its W_{neck} and $W_{str\text{-}hard}$ values are higher) because the lower σ_y value of the material allows the film's stretching zone to propagate farther under the falling dart (and to absorb more of the dart's energy) before the stress in the smallest cross section of the film reaches the breaking point, σ_{br}. A still further increase in the 1-butene content to 5.5 mol % decreases all the σ values even more, but the dart impact strength value remains approximately the same.

Table 5.7 Effect of Resin Density and Copolymer Composition on Mechanical Properties and Dart Impact Strength of 25-μm (1-mil) Film from Compositionally Nonuniform Ethylene/1-Butene Copolymers

Density, C_M^{copol} [a]	Mechanical properties[b]							Estimated dart impact strength[b]				
	σ_y	ε_y	σ_n	ε_n	σ_{br}	ε_{br}	m_{exp}	m_{calc}	W_{yield}	W_{elast}	W_{neck}	$W_{str\text{-}hard}$
0.921/3.5	1.32	1.14	1.07	5.7	1.91	8.9	~120	145	1	5	62	27
0.917/4.4	0.96	1.33	0.89	5.4	1.75	8.9	~280	240	4	12	87	56
0.910/5.5	0.64	1.53	0.60	5.8	1.35	8.9	~230	235	6	15	82	53

[a] Density, g/cm³/C_M^{copol}, mol %
[b] All σ values are in kg/mm², all W values are in kg · mm, both m_{dart} values are in g

The same effect is characteristic for film made from ethylene/1-hexene LLDPE resins. An increase of the 1-hexene content from 2.3 to 3.2 mol % and the respective decrease in density from 0.923 to 0.917 g/cm³ results in a significant decrease of all three σ values. This change leads to an increase of the calculated dart impact strength from 230 to 290 g.

5.2.6 Effect of Compositional Uniformity

LLDPE resins of high compositional uniformity are produced with single-site metallocene catalysts. Table 5.8 lists mechanical properties of several resins of this type and compares their estimated and experimentally measured dart impact strength. A comparison with the data in Table 5.6 and in Figure 5.6 shows that, overall, compositionally uniform resins are softer and are more easily stretched (their σ_y and σ_n values are lower). However, their necking range is usually shorter (ε_n ~ 4.5 rather than 5.5), and the strain-hardening range, $\varepsilon_{br} - \varepsilon_n$, is longer, indicating that resins of this type exhibit more expressed rubber-like properties in the oriented state.

The superior dart impact strength (both experimentally determined and predicted by the model) for film made from compositionally uniform resins is mostly attributed to their strain-hardening properties; the $W_{\text{str-hard}}$ term in the W_{total} value (Equation 5.9) for them is significantly higher than for film made from compositionally nonuniform LLDPE resins.

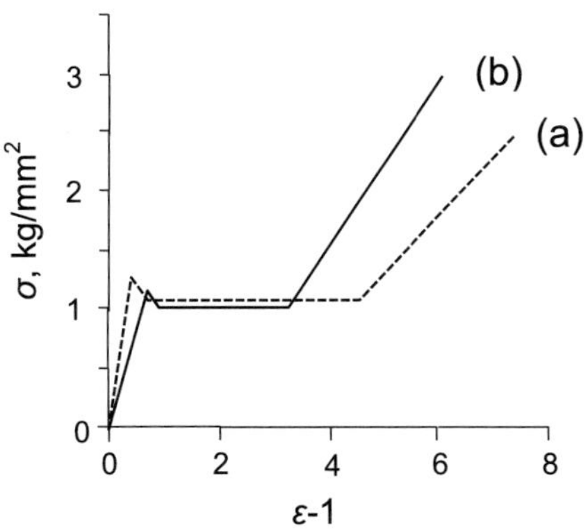

Figure 5.6
Stress/strain curves of ethylene/1-hexene copolymers of the same crystallinity degree (density): (a) compositionally nonuniform resin; (b) compositionally uniform resin

Table 5.8 Mechanical Properties and Dart Impact Strength of 25-μm (1-mil) Film from Compositionally Uniform Ethylene/α-Olefin Resins

Resin[a]	Mechanical properties[b]								Estimated dart impact strength[b]			
	σ_y	ε_y	σ_n	ε_n	σ_{br}	ε_{br}	m_{exp}	m_{calc}	W_{yield}	W_{elast}	W_{neck}	$W_{str\text{-}hard}$
E/H, 0.917	1.14	1.93	1.02	4.6	3.04	7.8	>800	830	33	57	263	197
E/H, 0.916	1.03	1.80	1.00	4.4	2.76	8.1	>800	720	26	44	200	204
E/MP, 0.914	0.96	1.46	0.89	4.3	2.33	7.6	>800	460	11	11	136	131

[a] E/H - ethylene/1-hexene copolymers, MP - ethylene/4-methyl-1-pentene copolymer; density in g/cm³
[b] All σ values are in kg/mm², all W values are in kg · mm, m_{dart} values are in g

5.3 Tear Strength of LLDPE and LDPE Film

The tear strength is the second important end-use mechanical parameter of blown and cast film manufactured from HDPE, LLDPE, and LDPE resins [1–4]. Commercial publications providing general data on properties of polyethylene resins usually contain the results of the tear strength test [5–7]. However, all this information is strictly empirical and provides practically no insight with regard to possible correlations between the resin structure and its basic mechanical properties on one side and this end-use film property on the other.

5.3.1 Description of Tear Test

The tear test is carried out according to ASTM Method D1922 and ISO Method 6383-2 on a special apparatus (Elmendorf apparatus) following the standard procedure [16]. Several rectangular pieces 76 mm (3 in) in width and 63 mm (2.5 in) in length are cut from a roll of film in two directions: the machine direction of the roll and the transverse direction. The principal component of the Elmendorf apparatus is a pendulum with the arm length of 40.6 cm (16 in). The weight of the pendulum can be varied from 200 to 1,600 g depending on the tear strength of the film. The pendulum is fixed in the horizontal position, and the long edge of the film is fastened by two clamps (see Figure 5.7(a)), one attached to the pendulum and another to the body of the apparatus. A cut is made with a razor in the middle of the wide side of the film strip. The cut leaves the undamaged length of the film equal to 43 mm (1.7 in). The end of the cut, the point where the tear starts, extends ~5 mm beyond outer edges of the clamps.

When the pendulum is released, its swing motion rapidly tears the uncut length of the film, as shown in Figure 5.7(b) and Figure 5.7(c). Due to the geometry of the Elmendorf apparatus, the film is completely torn when the pendulum reaches the lowest point of its swing. Without the film, the linear speed at this point would be the highest, ~1.2 m/sec (~47 in/sec), but the film-tearing process strongly slows down the pendulum. The duration of the single test is less than 0.5 second. The energy loss of the pendulum is determined by measuring its upswing after the film strip is torn, it is used to calculate the tear strength of the film. Test results for several pieces of the same film are averaged; the tear strength of film is defined as the weight of the pendulum, in gram, which is sufficient to bring the pendulum to a halt at the moment when the film is completely torn.

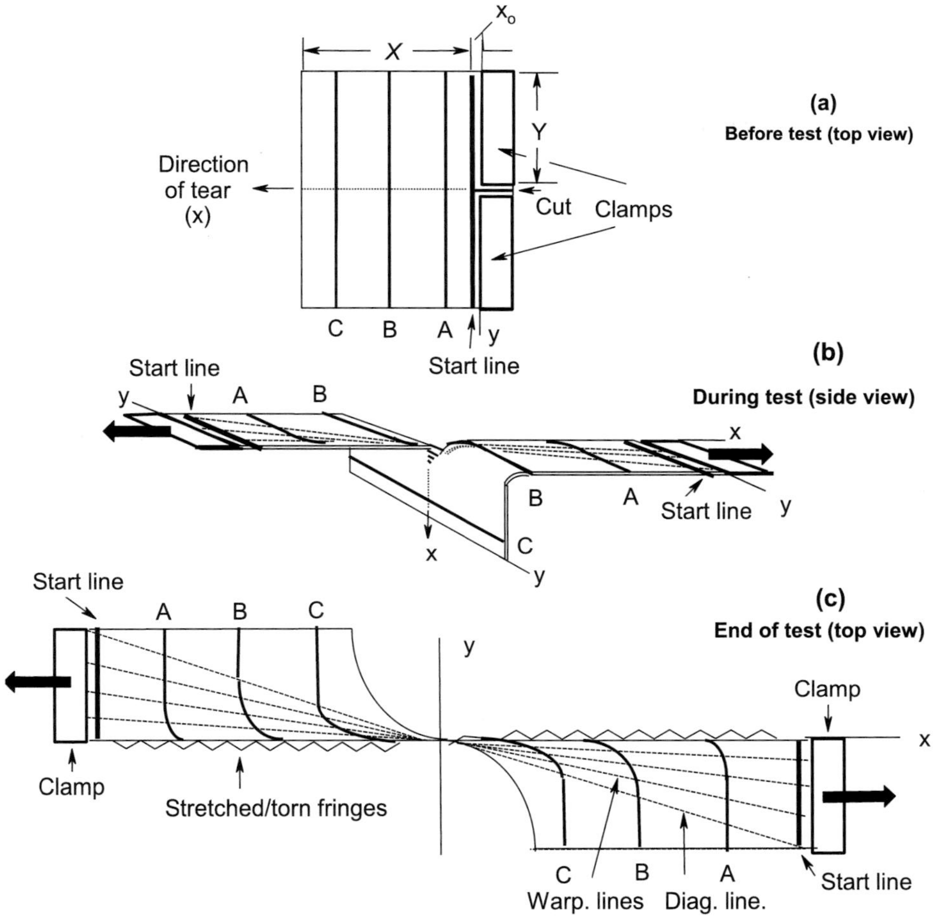

Figure 5.7 Schematics of Elmendorf tear test

5.3.2 Physical Details of Tear Test

The principal goal in the analysis of the tear process is to visualize all the steps occurring during the film tearing in the terms conventionally used to describe standardized mechanical testing of semi-crystalline plastics (Section 1.10). Mechanical processes that occur during the tear test can be seen best when high quality LLDPE film is used [17]. As Figure 5.7(b) shows, the tearing process is accompanied by a temporary twisting of the still undamaged part of the film, which is now positioned perpendicularly to the plane of the pendulum movement. The twisting results in an apparent turning of the *x* coordinate by ~ 90° whereas the direction of the *y* coordinate, which is perpendicular to the tear direction, remains unchanged. The original direction of the *x* coordinate is restored at the moment when the film is completely torn, as shown in Figure 5.7(c).

Tearing of a film strip during the test occurs very rapidly, and its direct observation is difficult. Only a few features are easily noticeable. The first one is the formation of border fringes; narrow, strongly stretched borders along the tear line, which are very schematically presented in Figure 5.7(c). Figure 5.8 shows the microphotograph of a typical fringe. The fringes are usually small; their lengths vary from ~ 0.3 to ~ 1 mm, depending on the type of film and tear direction.

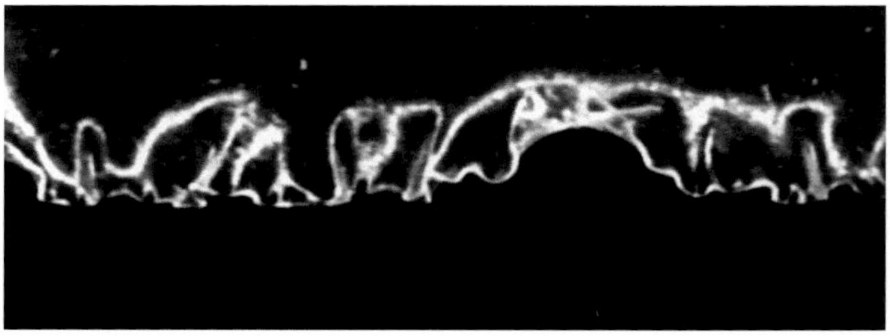

Figure 5.8 Microphotograph of oriented fringes in transverse-torn LLDPE film; magnification × 100, length of fringes ~ 0.7 mm

The second directly observable feature is straining of the area in the film between the *x* coordinate and the diagonal lines shown in Figure 5.7(b). For convenience of observation, three lines marked A, B, and C in Figure 5.7(a) were drawn on the film before the test perpendicular to the tear direction. The straining results in bending of these lines. The bending increases with the length of the tear (Figure 5.7(c)). The shape of the bent lines can be measured in torn film samples after the test.

When the tear test is performed on a film sample produced by the film-blowing method, a well-known aberration in the test might occur. Although the tear is supposed to propagate along the *x* coordinate, it often changes its path and ends up propagating in the *y* direction. The reason for the direction change is the large difference in the orientation pattern in the film. Some types of blown film are quite strongly machine- or transverse-oriented. As a result, their tear strength in the two directions is significantly different, and the tear direction may shift to follow the path of the least resistance.

Additional features of the tear test become noticeable if the tear is carried out at a low speed [17]. The first such feature is very strong warping of film regions along the tear line. The directions of the warping lines are shown in Figure 5.7(b) and Figure 5.7(c). The second feature is a high stress in some parts of the film during the test. The stress is present in the area between the tear line and the last warping line (the diagonal line in Figure 5.7(c)) while all other parts of the film are not stressed at all.

Figure 5.9 shows details of an interrupted tear test. Only the film segments strained during the test are shown. The tearing involves three parts of the film, Region I and two adjacent Regions II (Figure 5.9(a)). The central part, Region I, is affected most strongly. This region has the form of an isosceles triangle; its base is positioned along the x axis at $y = 0$, and the region is twisted with respect to the x coordinate. Region I is always small; its altitude is usually less than ~3 to 4 mm. A small part of Region I in the middle of the triangle base, the area about to be torn apart, is strongly and irreversibly stretched and under the breaking stress, σ_{br}.

Figure 5.9(b) shows schematically what happens with one half of Region I when the tear point continues to travel along the altitude of the triangle (along the x coordinate). At the moment when the tear point reaches the position shown at left in Figure 5.9(b), the height of Region I (its altitude) is h', the distance which can be approximately estimated experimentally, as shown below. As the tear point travels along the altitude toward the vertex of Region I, the altitude h' becomes slightly elongated and gradually splits into two halves, h'', which are positioned along the x coordinate. This change is shown in the right part of Figure 5.9(b), at a moment when the tear point has moved far ahead ($x_{tear}'' >> x_{tear}'$).

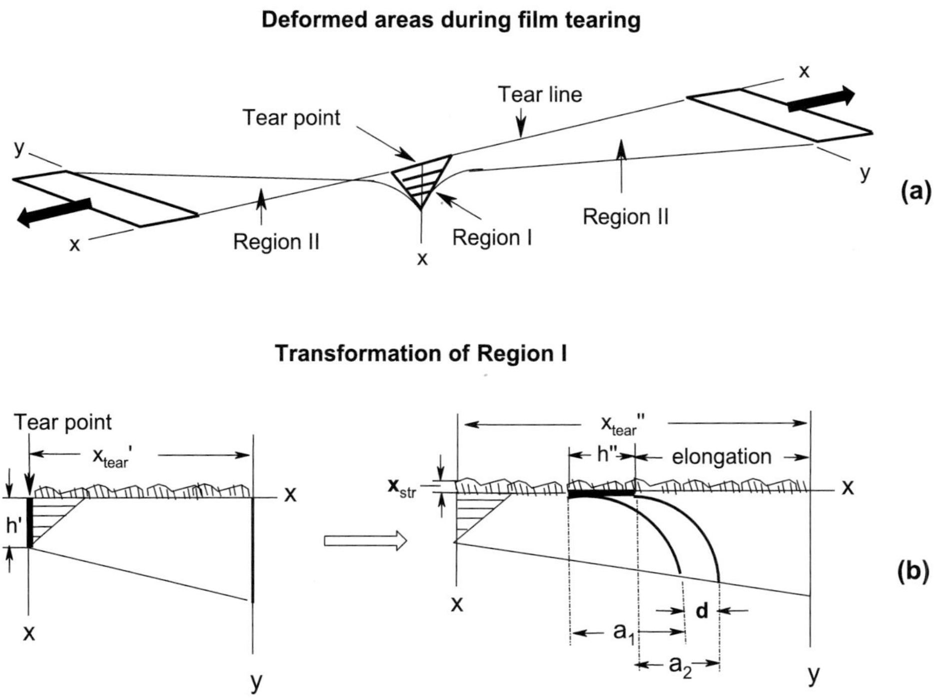

Figure 5.9 (a) Stressed/stretched film regions during tear test; (b) transformation of Region I during passing of tear point along its altitude

Several important visual observations become possible if such slow tearing is stopped midway and the sample is held under stress only slightly below σ_{br} [17]:

1. The size of Region I in the strained sample, both its base and altitude, change relatively little when the total length of the tear increases.
2. In several tests, a short distance along the x axis was marked (like h' in Figure 5.9(b)), and the tear was allowed to propagate along the marked distance. The measurement of its both halves (h'') showed that after the torn film was subsequently relaxed, h'' were only slightly larger than the initial h'.

These observations were confirmed in a more quantitative manner. Prior to testing, a large number of lines were drawn on the film strip perpendicular to the x coordinate, similar to lines A, B, and C in Figure 5.7(a), with short distances **d** between the lines. The tearing process transformed the end of each line into a curve (Figure 5.9(b)). After the tests, the lines remained curved and each ended with a short, strongly oriented fringe shown in Figure 5.8.

Measurements carried out with completely torn and relaxed film samples showed that the $a_1 - a_2$ distances are approximately proportional to h'' and that the $(a_1 - a_2)/\mathbf{d}$ ratios give the level of irreversible strain in the film along the tear line, ε_{film}. Figure 5.10 and Figure 5.11 show dependences for the total elongations along the tear line and the ε_{film} values for samples of blown film manufactured from LLDPE and LDPE and torn in the machine and the transverse direction. Figure 5.12 gives x_{fringe} values for the two types of film.

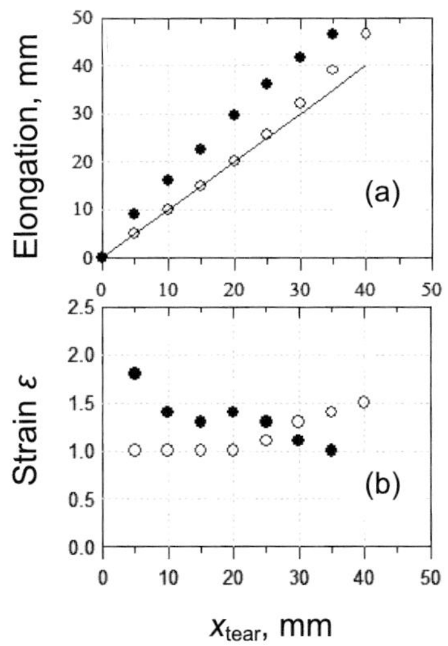

Figure 5.10
(a) Irreversible elongation along tear line (distances between x_{tear} = 0 and ends of curved lines) in relaxed LLDPE film torn in transverse (●) and machine (○) direction; (b) irreversible strain $(a_1 - a_2)/\mathbf{d}$ (see Figure 5.9(b)) in LLDPE film torn in transverse (●) and machine (○) direction

Figure 5.11 shows that the strain in the LLDPE film (ε_{film}) along the tear line depends on the tear direction. Tearing in the machine direction proceeds easily and without any noticeable strain except for the last stage before the complete breaking of the film strip. In contrast, tearing in the transverse direction results in the development of a significant permanent strain ($\varepsilon_{film} > \varepsilon_y$) in the beginning of the tearing process (at low x_{tear}); whereas the final stages of the tearing proceed without any noticeable additional straining. Because the film used in the transverse and the machine-direction tests is made from the same LLDPE resin, these results can only be reconciled if one takes into consideration that the polymer chains in blown HDPE and LLDPE film are always preferably oriented in the machine direction. (see Section 5.1.2) [14, 18, 19]. The data in Figure 5.10 show that even a small difference in the degree of chain orientation in LLDPE film significantly affects its tear behavior. The strain always significantly exceeds ε_y, which is most probably caused by rapid machine → transverse reorientation of polymer chains in the course of tearing.

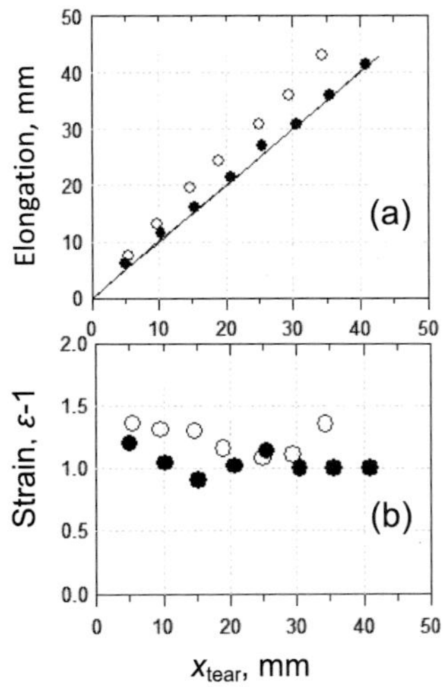

Figure 5.11
(a) Irreversible elongation along tear line (distances between x_{tear} = 0 and ends of curved lines) in relaxed LDPE film torn in transverse (●) and machine (○) direction; (b) irreversible strains $(a_1 - a_2)/\mathbf{d}$ (see Figure 5.9(b)) in LDPE film torn in transverse (●) and machine (○) direction

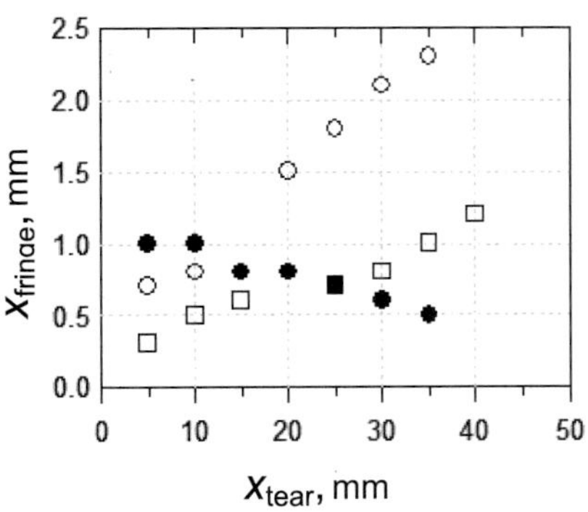

Figure 5.12
Length of stretched/oriented fringes x_{str} in LLDPE film torn in machine direction (□) and in LDPE film torn in transverse (●) and in machine (○) direction

Film made from LDPE resins has a very complex pattern of chain/crystal orientation [18, 19]. Figure 5.10 shows that the tear pattern in this film is reversed; the initial straining is slightly higher in the film torn in the machine direction, but overall, the straining during the tearing in both directions is relatively small and the ε_{film} value is close to the respective ε_y values. In general, the ε_{film} values for both types of film are within 1.0 and 1.3. This means that every time the tear point advances by a distance h' along the x coordinate, the length of the partially torn sample, $2x_{tear}$, increases by $2\ h'' \approx 2 \cdot \varepsilon_{film} \cdot h'$ (Figure 5.9(a)).

Figure 5.9(a) shows that two other areas of the film, Regions II, are also strained during the tear test. These are two orthogonal trapezoids; their large bases are equal to the width of the clamp, Y, and their small bases correspond to the height h' of Region I. The stresses and the strains at any point along the x coordinate in Regions II are very uneven and depend on the position along the y coordinate. The highest strains are along the tear line, at $y = 0$, but judging by the data in Figure 5.10 and Figure 5.11, all these strains are lower than ε_y. The yet undamaged part of the film in advance of the apex of Region I (Figure 5.9) does not experience any strain.

5.3.3 Model of Tear Test

Significant complications arise when an analytical solution is attempted to describe stretching of a combination of Region I and two adjacent Regions II (Figure 5.9). One reason for the difficulty is strong twisting and warping of all the stress-affected parts of the film, a fact obvious from Figure 5.7(b) and Figure 5.9(a). The second complication is a very uneven distribution of stresses and strains in Region I. Obviously, the center of the base in Region I is under the breaking stress. However, the overall strain in the base of the triangle is smaller than either σ_{br} or σ_n due to cohesion of the material in the base and the underlying sections of Region I. The cohesion level is different when the test is carried out in different directions of the film (Figure 5.10 and Figure 5.11).

The model of the tear test [17] divides a combination of Region I and the two adjacent Region IIs into a large number of primitive elements. Figure 5.13(a) and Figure 5.13(d) show the top primitive element from Figure 5.9(a) and its internal coordinates. The element consists of two narrow equilateral trapezoids fused at their small bases, dx_{tear}, which is the distance that corresponds to the elemental step along the tear line. The coordinate origin in the primitive element is positioned in its middle (the narrowest point); any distance from the origin is designated as ξ (Figure 5.13(d)). The minimum length of the first element at the beginning of the tearing process is $2x_0$ (10 mm or ~0.4 in), its length at any given x_{tear} is $2 \cdot (x_{tear} + x_0)$, and its maximum length is $2 \cdot (X + x_0)$ (~96 mm or ~3.8 in). Each primitive element is twisted 180° with respect to the x coordinate (Figure 5.10(a)); this twisting does not affect the mechanics of its straining.

During the test, the primitive element is gradually stretched until the stress at $\xi = 0$ (in the narrowest cross section, S_{min}) reaches the breaking stress, σ_{br}. At this moment, all the elements positioned below the top element in Figure 5.9(a) are under lower stresses, and the narrowest cross section of the last element, at the apex of Region I, reaches the yield point (σ_y and ε_y). After the two parts of the top primitive element are torn apart under the breaking stress (Figure 5.13(c)), Region I is reconstituted. The next primitive element becomes the most-stressed element; all underlying elements are additionally strained, and the bottom point of Region I propagates down the x coordinate by the distance dx_{tear}. In other words, *the tear test consists of the cascade of straining and breaking of the primitive elements* until x_{tear} reaches X and ξ reaches $X + x_0$.

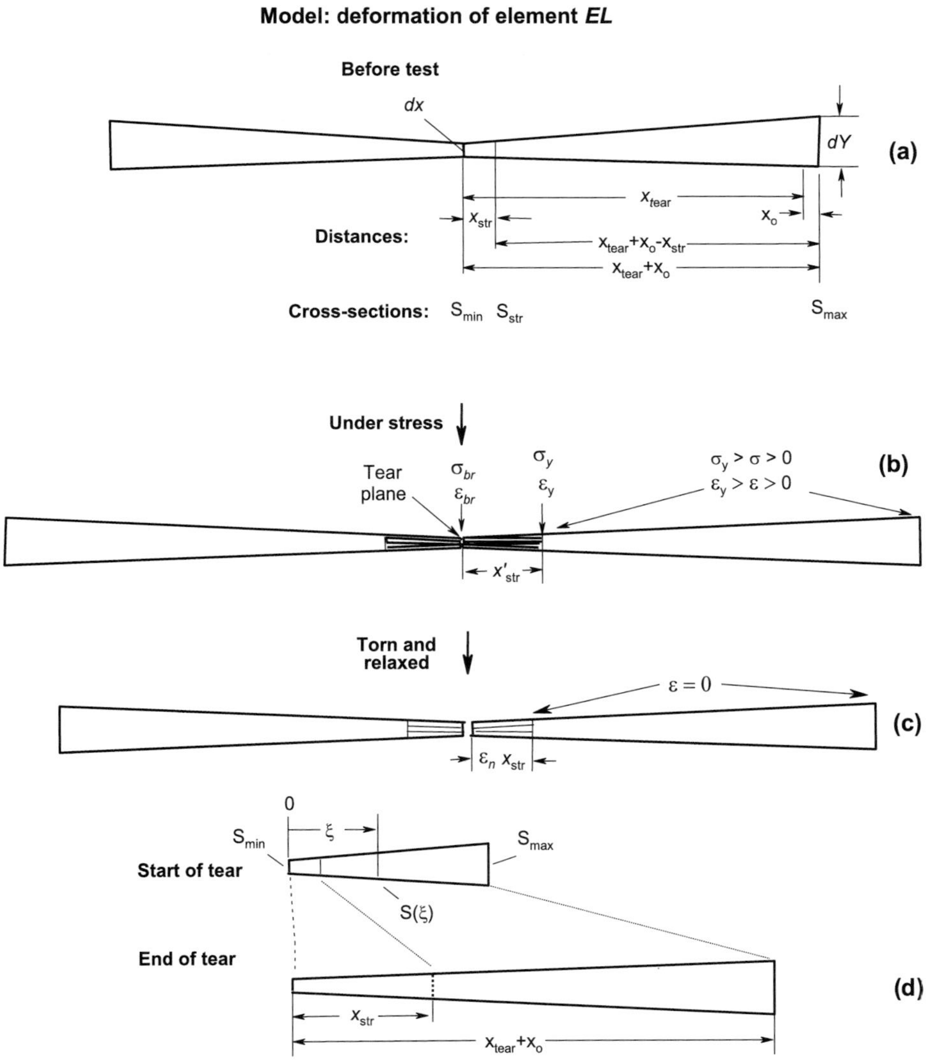

Figure 5.13 Deformation of primitive element: (a) before test; (b) during test, under stress; (c) after test, torn and relaxed; (d) internal coordinate and relationships between original and stretched parts in one half of primitive element

Figure 5.13 signifies that all the stages of stretching/straining of the primitive elements are in effect the stages of the standard tensile test of an object with a linearly variable cross section, the smallest in the middle ($S_{min} = t \cdot dx_{tear}$) and the largest at both ends, $S_{max} = t \cdot dY$, where t is the film thickness.

At the moment when the primitive element is about to break into two halves (Figure 5.13(b)), its total length substantially increases. First of all, the two irreversibly deformed parts, $2x_{str}$, lengthen to $2x_{fringe}$. The deformation in both strained x'_{str} parts varies from the breaking strain ε_{br} in the smallest cross section S_{min} to the yielding strain ε_y at their opposite ends, S_{str}. The remaining parts of the element, each with the length $x_{tear} + x_0 - x_{str}$, are elastically strained below the yield point, and their strain gradually decreases starting from ε_y. When the primitive element is torn and relaxed (Figure 5.13(c)), both its irreversibly stretched parts at $x_{fringe} = \varepsilon_n \cdot x_{str}$ appear as fringes on the torn halves of the film (Figure 5.9(b), and Figure 5.12). In principle, the previously elastically strained parts of the element should contract to their unstrained combined length, $2 \cdot (x_{tear} + x_0 - x_{str})$. However, polyethylene resins do not have sharp clear-cut yield points and these distances in the relaxed film can be noticeably higher.

The goal of the model is to describe the shape of the primitive element in Figure 5.13 at different points of the tear propagation path along the x coordinate and calculate the total work of tearing each primitive element. After the tear test is completed, the energy loss of the pendulum is equal to the work of the gradual tearing of a piece of film (one primitive element after another), W_{total}. By definition of the test, the tear strength (tear resistance) of the film is equal to the potential energy of the pendulum with the weight m_{tear} (g) released at the 90° angle and stopped at the lowest point of its swing due to the film's resistance to tear:

$$W_{total} = 2 \cdot m_{tear} \cdot g \cdot l \cdot \sin^2(90°/2) = m_{tear} \cdot g \cdot l \ \ (\text{g} \cdot \text{m}^2/\text{sec}) \tag{5.14}$$

Here l is the pendulum length, 40.6 cm (16 in), and $g = 9.81$ m/sec^2.

Similar to the model of the dart impact test described in Section 5.2, the total work required to strain/stretch and break all the primitive elements, W_{total} in Equation 5.14, is separated into four parts:

$$W_{total} = W_{yield} + W_{neck} + W_{str\text{-}hard} + W_{elast} \tag{5.15}$$

The first three terms in Equation 5.15 describe the work spent on irreversible deformation of the parts of all the primitive elements from $\xi = 0$ to $\xi = x_{str}$ and the last term describes the work of reversible elastic stretching in cross sections of all the primitive elements beyond $\xi = x_{str}$ (Figure 5.13):

1. W_{yield}, the work of elastic stretching in cross sections of all the primitive elements from $\xi = 0$ to $\xi = x_{str}$ to the yield point;

2. W_{neck}, the work of necking, that is, stretching/orientation of the film material from $\xi = 0$ to $\xi = x_{str}$;
3. $W_{str\text{-}hard}$, the work of strain-hardening, additional straining of oriented parts in all the elements to the point at which the stress in the smallest cross sections reaches σ_{br}.
4. W_{elast}, the work of sub-yield elastic stretching in cross sections of all the primitive elements beyond $\xi = x_{str}$ up to $\xi = x_{tear} + x_0 - x_{str}$.

Straining of a primitive element proceeds along the ξ coordinate, with the origin positioned in the middle of the element. Cross-section areas $S(\xi)$ of the element at a distance ξ from the origin are given by Equation 5.16 [17]:

$$S(\xi) = [t \cdot dx_{tear}] \cdot [1 + \xi \cdot (dY/dx - 1)/(x_{tear} + x_0)] \tag{5.16}$$

For every such cross-section area, the work of deformation $W_{i,\xi}$ is the product of the force $F_{i,\xi} = \sigma_i \cdot S(\xi)$ and the distance of the force application, $d\xi$, from ξ to $\xi + d\xi$.

Although analytical forms for all the terms of W_{total} in Equation 5.15, (W_{yield} [total], W_{neck} [total], $W_{str\text{-}hard}$ [total] and W_{elast} [total]) are desirable, some final expressions, after two integration steps, are very cumbersome due to the complex shape of the primitive element. These calculations can be carried out only by numerical integration of respective differential equations.

Calculation of W_{yield}:

$$W_{yield}(\text{element}) = \sigma_y \cdot (\varepsilon_y - 1) \cdot [t \cdot dx_{tear}] \cdot \left\{x_{str} + [r_{tear}/(\sigma_{br}/\sigma_y - 1) - 1] \cdot 0.5 \cdot x_{str}^2/(x_{tear} + x_0)\right\} \tag{5.17}$$

$$W_{yield}(\text{total}) = \int_{x_{tear}}^{X} W_{yield}(\text{element}) \cdot dx_{tear} \tag{5.18}$$

Calculation of W_{neck}:

$$W_{neck}(\text{element}) = 2 \cdot \sigma_{br} \cdot (\varepsilon_n - \varepsilon_y) \cdot t \cdot r_{tear} \cdot (x_{tear} + x_0) \cdot dx_{tear} \tag{5.19}$$

$$W_{neck}(\text{total}) = \int_{x_{tear}=0}^{X} W_{neck}(\text{element}) \cdot dx_{tear} =$$

$$= 2 \cdot \sigma_{br} \cdot (\varepsilon_n - \varepsilon_y) \cdot t \cdot r_{tear} \cdot X \cdot (0.5 \cdot X + x_0) \tag{5.20}$$

Calculation of $W_{str\text{-}hard}$:

$$W_{str\text{-}hard}(\text{element}) = 2 \cdot \sigma_{br} \cdot [t \cdot dx_{tear}]^2 \cdot (\varepsilon_{br} - \varepsilon_n) \cdot \int S(\xi)^{-1} \cdot d\xi \tag{5.21}$$

$$W_{\text{str-hard}}(\text{total}) = \int_{x_{\text{tear}}=0}^{X} W_{\text{str-hard}}(\text{element}) \cdot dx_{\text{tear}} \tag{5.22}$$

Calculation of W_{elast}:

$$W_{\text{elast}}(\text{element}) = \int_{\xi=x_{\text{str}}}^{x_{\text{tear}}+x_0-x_{\text{str}}} (\sigma_{br}^2/\sigma_y) \cdot (\varepsilon_y - 1) \cdot \left[S_{\min}^2/S(\xi)\right] \cdot d\xi \tag{5.23}$$

$$W_{\text{elast}}(\text{total}) = \int_{x_{tear}=0}^{X} W_{\text{elast}}(\text{element}) \cdot dx_{\text{tear}} \tag{5.24}$$

The model defined by Equation 5.17 to Equation 5.24 has only one adjustable parameter, the ratio $r_{\text{tear}} = x_{\text{str}}/(x_{\text{tear}} + x_0)$. Estimations of r_{tear} in Figure 5.12 give its range from 0.06 to 0.08.

This model of the tear test and the numerical calculations with Equation 5.15 to Equation 5.24 reveal several important features of the test and afford the estimation of the effects of resins' mechanical properties on the tear strength of film. As a test of the model reliability, the following experimentally determined parameters typical for a commercial LLDPE resin were used to estimate the average tear strength of film:

σ_y = 1.3 kg/mm^2, ε_y = 1.2, σ_n = 1.2 kg/mm^2, ε_n = 6.0, σ_{br} = 2.5 kg/mm^2, ε_{br} = 7.5.

Calculations for 25-µm (1-mil) film prepared from such a resin without account for the orientation effect give the tear strength in the range of 170 to 220 g, a correct order of magnitude of the tear strength when compared to the experimental data (see below). The energy of the pendulum is spent in the following way:

1. Stretching the narrow part of each primitive element to the yield point, W_{yield} = 1.1%;
2. Necking/orientation of this part of the element, W_{neck} = 73.4%;
3. Overcoming strain hardening in the element, $W_{\text{str-hard}}$ = 20.0%;
4. Elastic deformation of the nonoriented part of the element, W_{elast} = 5.5%.

It is very informative to compare two calculated values for a strip of film:

a) The work of tearing the piece of the film in the tear test, which is essentially a modified strain/stress test in the course of which narrow strands of film break one after another, and

b) The work of stretching and breaking of a rectangular piece of the same film with the cross section $S_{\max} = t \cdot X$ and the length $2Y$.

The calculations show that the total work of breaking the film in the tear test, $W_{\text{total}} = 90$ kg · mm, constitutes merely 11% of the work required to break the same piece of film when it is tested in the standard strain/stress mechanical test.

5.3.4 Effect of Pendulum Speed

Some mechanical parameters of polyethylene resins depend on the deformation rate, mostly due to the relatively slow relaxation phenomena described in Section 5.1.2. The necking stage is the crucial part of film tearing, and only the effect of the tearing speed V on σ_n and ε_n values was considered in the tear test model. The same resin, when stretched at a high speed typical for the tear test, is slightly stiffer during the necking stage, and its final necking strain is noticeably higher. These two effects increase the calculated tear resistance. For example, the estimated tear strength for the film produced from the LLDPE resin with the above set of σ and ε values increases from ~220 g at a low stretching speed to ~260 g at a high stretching speed.

5.3.5 Effect of Resin Mechanical Properties

All six parameters that determine the shape of the stress/strain curve, σ_y, ε_y, σ_n, ε_n, σ_{br}, and ε_{br}, affect the tear strength of film in a complex way due to their contribution to different terms in the expression for the total work of film tearing, W_{total} in Equation 5.15. Some computational results for 25-µm (1-mil) film are presented in Table 5.9. They show the predicted effects of mechanical properties of LLDPE resins on the calculated tear strength of film. A pronounced yield threshold, $\sigma_{br}/\sigma_y > 1$, which is typical for some LLDPE resins of higher density and crystallinity degree, does not affect the tear strength of the film. The calculations show also that the value of the Young's modulus, the $M_Y = \sigma_y/(\varepsilon_y - 1)$ ratio, has virtually no effect on the tear resistance. An increase of resistance to orientation (a proportional increase of σ_y and σ_n values) also produces only a modest increase of the tear strength, whereas an increase of the orientation (necking) potential (an increase in ε_n at constant $\varepsilon_{br} - \varepsilon_n$) significantly increases the tear strength. An increase in the strain-hardening modulus of the stretched polymer, $M_{\text{str-hard}} = (\sigma_{br} - \sigma_n)/(\varepsilon_{br} - \varepsilon_n)$, also strongly increases the film's tear strength by allowing the orientation zone in each primitive element to propagate farther. Finally, an improvement in strain-hardening properties of a resin (a parallel increase of σ_{br} and ε_{br} at a constant $M_{\text{str-hard}}$) also greatly increases the tear strength of the film.

Table 5.9 Effects of LLDPE Mechanical Properties on Film Tear Strength

Yield threshold, $\sigma_y/\sigma_n = 1.2$ ***kg/mm²,*** $\sigma_{br} = 3.0$ ***kg/mm²,*** $\varepsilon_y = 1.2$, $\varepsilon_n = 6$, $\varepsilon_{br} = 8$					
σ_y/σ_n:	1.02	1.05	1.10	1.15	
Tear strength m_{tear}, g:	267	258	270	271	
Resistance to orientation; proportional increase of yield stress and necking stress at $\sigma_y/\sigma_n = 1.1$					
σ_n, kg/mm²:	0.8	1.0	1.2	1.4	1.6
Tear strength m_{tear}, g:	259	264	270	276	281
Increase of orientation potential ($\varepsilon_{br} = \varepsilon_n + 2.0$)					
ε_n:	4.0	6.0	8.0		
Tear strength m_{tear}, g:	200	270	340		
Increase of strain hardening modulus $M_{str\text{-}hard} = (\sigma_{br} - \sigma_n)/(\varepsilon_{br} - \varepsilon_n)$; $\sigma_n = 1.2$ ***kg/mm²***					
$M_{str\text{-}hard}$, kg/mm²:	1.0	1.5	2.0		
Tear strength m_{tear}, g:	285	360	440		
Increase of strain hardening region ($\sigma_n = 1.2$ ***kg/mm²,*** $M_{str\text{-}hard} = 1.0$)					
$\varepsilon_{br} - \varepsilon_n$:	2.0	3.0	4.0		
Tear strength m_{tear}, g:	285	390	500		

5.3.6 Effect of Film Orientation

Large differences between the tear strength in the machine and the transverse direction in blown LLDPE film is one of the most well-known features revealed in tear tests. Two examples of the difference are presented in Table 5.10. This difference is the consequence of a significant chain orientation in LLDPE film.

Table 5.10 Effect of Orientation on Mechanical Properties and Tear Strength of Blown Film from Ethylene/1-Hexene LLDPE Resins

Density, g/cm³	Direction of test	Mechanical properties						Tear strength m_{tear}, g	
		σ_y, kg/mm²	ε_y	σ_n, kg/mm²	ε_n	σ_{br}, kg/mm²	ε_{br}	Calculated	Experiment[a]
0.918	Machine	1.6	1.1	1.6	3.3	6.1	6.8	~370	300
	Transverse	1.5	1.1	1.4	7.1	4.8	9.0	~470	670
0.917	Machine	0.9	1.1	0.9	3.1	5.8	6.8	~320	440
	Transverse	1.0	1.1	0.9	7.2	4.8	9.4	~470	740
0.917	Machine	0.9	1.1	0.9	3.9	5.3	7.3	~340	460
	Transverse	0.9	1.1	0.9	6.9	5.1	9.2	~480	600

[a] Experimental data adjusted to film thickness 25 μm (1 mil)

Two features of the orientation pattern in the film are important for its tear strength. First, the *c* axis in the crystalline phase of polyethylene (the axis that corresponds to the direction of extended polymer segments in lamellae) in blown HDPE and LLDPE film is always preferably oriented in the machine direction. Table 5.1 gives several examples of the orientation pattern in typical LLDPE film. Second, the fraction of polymer chains in the crystalline phase oriented in the thickness direction of the film is always low, < 20% rather than 33% as expected in the absence of orientation. Consequently, the averaged mechanical strength of blown film is higher than the strength of an isotropic sample of the same resin.

This difference in orientation results in noticeable differences in the stretching process and in mechanical properties between the machine and the transverse direction of film. For example, estimations of the tear resistance corresponding to mechanical properties of the film shown in Figure 5.2 give the following results:

Tear strength in the machine direction ~ 290 g;

W_{yield} ~ 1%, W_{neck} ~ 48%, $W_{str\text{-}hard}$ ~ 45%, W_{elast} ~ 6%.

Tear strength in the transverse direction ~ 380 g;

W_{yield} ~ 1%, W_{neck} ~ 77%, $W_{str\text{-}hard}$ ~ 19%, W_{elast} ~ 3%.

The model of the tear test correctly predicts a higher tear resistance in the transverse direction for the film and explains the difference as mostly due to a significantly higher ε_n value in this direction.

Table 5.11 contains several examples in which the mechanical properties and the tear resistance of blown ethylene/1-hexene LLDPE film are experimentally measured in the machine and the transverse direction. A comparison between the experimentally measured tear strength and the estimations shows that the model correctly predicts the range of the tear resistance and differences between the tear properties in the two directions of film.

Table 5.11 Comparison of Mechanical Properties and Tear Strength of Blown Film Made from Two LLDPE Resins (Density 0.918 g/cm^3, I_2 = 1.0)

Copolymer	Direction of test	Mechanical properties						Tear strength m_{tear}, g 50-μm (2-mil) film	
		σ_y, kg/mm^2	ε_y	σ_n, kg/mm^2	ε_n	σ_{br}, kg/mm^2	ε_{br}	Calculated	Experiment
Ethylene/1-butene	machine	1.1	1.1	1.1	3.4	3.5	6.8	~460	350
	transverse	1.1	1.1	1.0	6.4	2.8	8.3	~520	590
Ethylene/1-hexene	machine	1.2	1.1	1.2	4.7	4.1	8.0	~690	820
	transverse	1.1	1.1	1.0	7.5	4.7	9.5	~950	1200

5.3.7 Comparison of Tear Strength of Ethylene/1-Butene and Ethylene/1-Hexene Copolymers

Film manufactured from ethylene/1-butene LLDPE resins is inferior in tear resistance to film manufactured from ethylene/1-hexene LLDPE resins of the same composition and density (crystallinity level), as the examples in Table 5.12 show [2, 4–7, 15]. The principal difference between these two types of resins is a lower strain-hardening modulus $M_{str\text{-}hard}$ for ethylene/1-butene copolymers [9]. Table 5.12 compares mechanical properties and the tear strength of the two types of blown film. Both resins were prepared using the same polymerization catalyst; they have the same crystallinity degree (density 0.918 g/cm^3) and molecular weight (I_2 = 1.0) [6]. The calculations show that the inferior tear strength of the film made from the ethylene/1-butene resin is indeed mostly the consequence of its lower σ_{br} value.

Table 5.12 Mechanical Properties of Two Types of LLDPE Film [5]

Property	Ethylene/1-butene copolymer		Ethylene/1-hexene copolymer	
	MD[a]	TD[a]	MD[a]	TD[a]
Yield strength σ_y, kg/mm^2:	0.96	0.97	0.86	0.79
Breaking strength σ_{br}, kg/mm^2:	5.42	3.60	6.74	3.90
Elongation ratio at break ε_{br}:	5.80	~8.5	3.43	~8.0
Dart drop strength m_{dart}, g:	100	81		
Elmendorf tear strength m_{tear}, g:	80	400	300	750

[a] MD – machine direction, TD – transverse direction

5.3.8 Low Density Polyethylene

LDPE resins are produced in radical homopolymerization reactions of ethylene at a high pressure. These polymers contain both short-chain branches of a different length (methyl, ethyl, and so on) and long-chain branches. These polymers have a reduced crystallinity level compared to HDPE resins prepared in catalytic homopolymerization reactions of ethylene, and they have lower melting points, 110 to 120 °C compared to ~135 °C for HDPE resins.

Table 5.13 lists mechanical properties of two commercial LDPE resins. A comparison with Table 5.11 and Table 5.12 shows that most mechanical properties of LDPE resins are similar to those of catalytically produced LLDPE resins except for a noticeably lower σ_{br} value. Machine and transverse orientation levels in LDPE film are relatively close [18, 19]. As a result, the experimentally measured tear strength in both directions of the LDPE film is also similar.

Calculations for 25-µm (1-mil) film support these results and demonstrate that the main reason for a relatively low tear strength of LDPE film compared to LLDPE film is lower σ_{br} and σ_n values and the resulting lower W_{neck} and $W_{str\text{-}hard}$ values:

LDPE (machine direction) $W_{total} \sim 85$ kg · mm

$W_{yield} < 1$, $W_{neck} \sim 58$, $W_{str\text{-}hard} \sim 17$, $W_{elast} \sim 8$ kg · mm

LLDPE (machine direction) $W_{total} \sim 140$ kg · mm

$W_{yield} < 1$, $W_{neck} \sim 110$, $W_{str\text{-}hard} \sim 25$, $W_{elast} \sim 3$ kg · mm

Table 5.13 Mechanical Properties and Estimated Tear Strength of LDPE Resins

Resin; density, g/cm³	Direction of test	Mechanical properties						Tear strength m_{tear}, g, for 50-µm (2-mil) film	
		σ_y, kg/mm²	ε_y	σ_n, kg/mm²	ε_n	σ_{br}, kg/mm²	ε_{br}	Calculated	Experiment
LD100 [6] 0.923	Machine	1.1	1.3	1.0	3.0	2.5	4.3	~260	150
	Transverse	1.1	1.3	1.0	4.5	2.2	6.5	~340	120
LDPE 132I [7] 0.921	Machine	1.2	1.3	1.1	4.0	3.2	5.7	~410	300
	Transverse	1.2	1.3	1.1	5.3	2.9	7.6	~500	180

5.4 Comparison of Factors Determining Tear Strength and Dart Impact Strength of LLDPE Film

The dart impact test and the tear test are the two most important end-use mechanical tests of film manufactured from LLDPE and LDPE resins. The models of the dart impact test (Section 5.2) and the tear test (Section 5.3) provide the basis for the comparison of both tests in terms of mechanical parameters of semi-crystalline resins; that is, the stresses and the strains at the yield, the end-of-necking, and the breaking points.

The models show that, formally speaking, the standard mechanical test of a film sample (such as ASTM D882, ISO 257-3), the dart impact test (ASTM D1709, ISO 7765), and the Elmendorf tear test (ASTM D1922, ISO 6383-2) are the variations of the same mechanical test but that they greatly differ in the testing conditions:

1. The standard stress/strain test is carried out with a strip of film of a constant cross section and at a relatively low stretching speed.
2. The dart impact test is carried out at a high stretching speed; its conditions correspond to stretching of a resin sample with a linearly varying cross section, see Figure 5.5.
3. The tear test is in effect a series if cascading high-speed tests of very narrow strips of film with variable cross sections and different lengths, see Figure 5.13.

Table 5.14 compares the work of film breaking in these three types of mechanical tests. The results were calculated for 25-μm (1-mil) film made from a hypothetical high-quality LLDPE resin with the following mechanical properties:

$$\sigma_y = 1.3\ \text{kg/mm}^2,\ \varepsilon_y = 1.2,\ \sigma_n = 1.2\ \text{kg/mm}^2,\ \varepsilon_n = 6.0,\ \sigma_{br} = 3.0\ \text{kg/mm}^2,\ \varepsilon_{br} = 8.0$$

The comparison shows that, in principle, the dart impact test and the tear test measure the same properties of the film but on a different scale because of major differences in the implementation of the respective tests. The only noticeable difference between both of the end-use tests and the standard stress/strain test (apart from elastic straining of the nonoriented part of the film, which is not accounted for in the standard test) is a higher contribution of the work of strain hardening in the standard stress/strain test. This difference is easily explained:

- The strain-hardening stage in the standard stress/strain test involves the whole length of the sample (Figure 1.18).
- The strain-hardening stage in the tear test is restricted only to two narrowest parts of the film adjacent to the tearing line (Figure 5.7 and Figure 5.8).
- The strain-hardening stage in in the dart impact test is restricted only to the narrowest cross section of the stretched film adjacent to the undamaged part of film clinging to the dart head (Figure 5.3)

Table 5.14 Comparison of Results of Three Different Mechanical Tests for LLDPE Film with Parameters: $\sigma_y = 1.3$, $\sigma_n = 1.2$, $\sigma_{br} = 3.0$ (all in kg/mm²); $\varepsilon_y = 1.2$, $\varepsilon_n = 6.0$, $\varepsilon_{br} = 8.0$

Test	W_{total}, kg · mm	W_{yield}, %	W_{neck}, %	$W_{str\text{-}hard}$, %	W_{elast}, %
Standard stress/strain test (cross section 1.0 mm²)	1,088	1.2	65.2	33.7	-
Tear test (total cross section 1.09 mm²)	109	0.9	75.9	18.1	5.0
Dart drop test (minimum cross section 0.58 mm²)	350	1.7	69.3	25.5	3.5

References

1. James, D. E., In *Encyclopedia of Polymer Science and Engineering*, vol. 6, 2nd ed. Kroschwitz, J. I. (Ed.) (1988) John Wiley & Sons, New York, p. 429
2. Krentsel, B. A., Kissin, Y. V., Kleiner, V. I., Stotskaya, S. S., *Polymers and Copolymers of Higher α-Olefins* (1997) Hanser, Munich, Chapter 8
3. *Ethylene Polymers, HDPE; Ethylene Polymers, LLDPE; Ethylene Polymers, LDPE*, In *Encyclopedia of Polymer Science and Technology*, vol. 2, 3rd ed. Mark, H. F., Kroschwitz, J. I., (Eds.) (2003) Wiley-Interscience, New York
4. Kissin, Y. V., In *Kirk-Othmer Encyclopedia of Chemical Technology*, vol. 17 (1996) John Wiley & Sons, New York, p. 756
5. ExxonMobil Chemical Company: *http://www.exxonmobilchemical.com/Chem-English/brands/polyethylene-lldpe-exxonmobil-grades-and-datasheets.aspx?ln = productsservices*
6. Dow Chemical Company: *http://www.dow.com/polyethylene/na/en/prod/index.htm*
7. Lyondell Company: *www.lyondell.com/Lyondell/Products/polymers/LinearLowDensityPolythylene*
8. Strella, S., In *Engineering Design for Plastics.* Baer, E. (Ed.) (1964) Reinhold, New York, Chapter 12
9. Kissin, Y. V., *Macromol. Mater. Eng.* (2008) 293, p. 66
10. Ward, I. M., *Polym. Eng. Sci.* (1984) 24, p. 724
11. Zachariades, A. E., Porter, R. S., *J. Macromol. Sci., Part B: Phys.* (1981) 19, p. 377
12. Tonkopij, E. A., Friedman, M. L., Kissin, Y. V., Malkin, A. Y., Zharova, I. L., Gul, V. E., *Plast. Massy* (1975) 10, p. 56
13. Standard Test Methods for Impact Resistance of Plastic Film by the Free-Falling Dart Method, ASTM D1709-04 09 A(ISO 7765-1); *http://www.astm.org/Standards/D1709.htm*, and D4272-09 (ISO 7765-2), Standard Test Method for Total Energy Impact of Plastic Films by Dart Drop, *http://www.astm.org/Standards/D4272.htm*
14. Kissin, Y. V., *J. Polym. Sci, Part B: Polym. Phys.* (1992) 30, p. 1165
15. Crotty, V. J., Firdaus, V., Hagerty, R. O., *SPE Polyolefins VIII International Conference.* (1993) Houston, TX, p. 192
16. Elmendorf tear test. Propagation Tear Resistance of Plastic Film and Thin Sheeting by Pendulum Method, ASTM D1922 (ISO 6383-2); *www.ptli.com/testlopedia/tests/ElmendorfD1922.asp*
17. Kissin, Y. V., *Macromol. Mater. Eng.* (2011) 296, p. 729
18. Krishnaswamy, R. K., *J. Polym. Sci, Part B: Polym. Phys.* (2000) 38, p. 182
19. Pazur, R. J., Prud'homme, R. E., *Macromolecules* (1996) 29, p. 119

6 End-Use Testing of High Molecular Weight HDPE and MDPE Resins

High density polyethylene (HDPE) and medium density polyethylene (MDPE) resins of high molecular weight are widely used for the manufacture of household and industrial containers, pipes, and tubes. This chapter describes the physical underpinning of several standard tests of various articles manufactured from these resins.

6.1 Top Load Test of HDPE Containers

Top load testing (crush testing) is the measurement of the resistance of a household container (usually a bottle) or an industrial container to a compressive load. The test determines the compressing load under which a container is irreversibly deformed or collapses. The compressive strength of bottles and cans made of polyethylene resins must be high enough to withstand the capping process of the containers, their stacking, storage, and shipment. The majority of household and industrial containers are single-use articles, and special government regulations for environmental protection were developed to limit the amount of resin spent for their manufacture.

Several standard procedures for the top load/crush tests are practiced [1–4]. The two main procedures are the dynamic test and the static top load test. In the dynamic test, a container is placed between two platens and the top movable platen compresses the container at a constant speed. The container is crushed until it collapses. Several parameters are recorded during the test: the breaking force (load) and the downward deflection of the top platen at this point, the yield force (the beginning of irreversible deformation), and the crushing force at a certain distance of the platen travel, usually between 12 and 25 mm (0.5 to 1 in). Such tests provide information for optimization of the force leading to deformation and failure of a container and for identification of excessively thinned areas in the wall of the container resulting from a poor wall-thickness distribution.

The need for the static top load testing is determined by the container storage conditions. Containers with packaged liquid products (water, beverages, milk, liquid detergents, etc.) are often stacked for significant periods of time. The bottom container in such a stack is continually stressed under a constant load of liquid-filled containers placed above it. In the static test, such containers, either empty or filled with liquid, are subjected to a predetermined static load. The test measures the crushing yield load (the load under which an increase in a downward deflection occurs even without any further increase of the load) and the crushing deflection, a decrease in the apparent height of the container at the crushing load.

6.1.1 Mechanics of Top Load Test

Figure 6.1(a) shows the schematics of a typical top load test. The tested container is usually a thin-walled bottle of the simplest configuration with a cylindrical body, a short, relatively thick neck (the thread on the neck is not shown), and a conical transitional section between the neck and the cylindrical body. The essential dimensions of the container are diameters D_1 and D_2; wall thicknesses t_1, t_2, t_3, and t_4; the height of the conical section h_1; and the height of the cylindrical part of the bottle H. The height of the neck (h_2) is irrelevant for the test results.

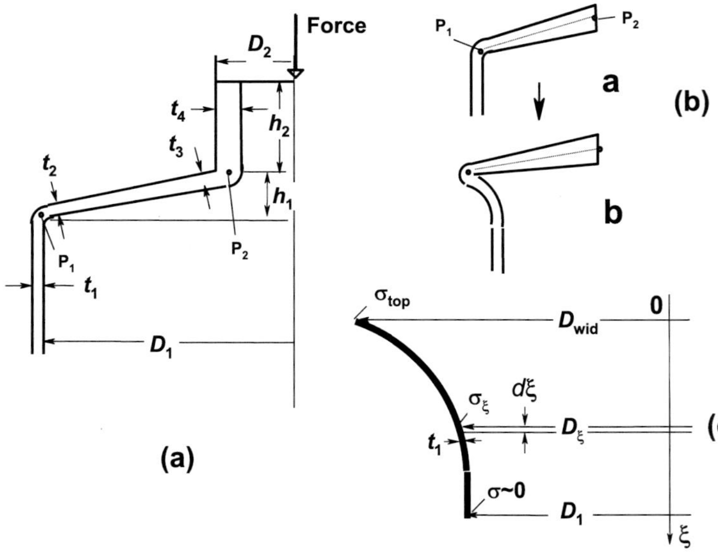

Figure 6.1 Schematics of top load test; (a) dimensions before test; (b) moment of initial failure; (c) stresses in model, Equation 6.3

The goal of the analysis presented below is to describe mechanical changes taking place during the dynamic top load test in terms of the standard stress/strain test of a polymer sample described in Section 1.10 of Chapter 1 (Figure 1.18).

The crushing force exerted by the top platen is uniformly applied to the top of the neck and the bottle is compressed until it collapses. Visual observations show that if $t_1 < t_2 < t_3$ (a typical case), one can neglect the downward bending of the transitional section of the bottle and assume that when the top load increases this section moves as a rigid pivot, which rotates at points P_1 and P_2. Figure 6.1(b) schematically shows the initial part of the test in an exaggerated form as the **a** → **b** transition. The downward movement of the top platen, which corresponds to the downward deflection of the neck, *def*, leads to an increase in the diameter of the upper edge of the cylindrical section. The diameter of the widened edge of the cylindrical body, D_{wid}, is:

$$D_{wid} = 2 \cdot [h_1^2 + 0.25 \cdot (D_1 - D_2)^2 - (h_1 - def)^2]^{0.5} + D_2 \tag{6.1}$$

This widening leads to stretching of the circumference at the top of the cylindrical body from $\pi \cdot D_1$ to $\pi \cdot D_{wid}$. When the $(\pi \cdot D_{wid})/(\pi \cdot D_1)$ ratio approaches the value of the yield strain ε_y, the upper part of the cylindrical body enters the zone of inelastic deformation on the stress/strain curve (Figure 1.18). To evaluate the maximum *reversible* downward deflection of the bottle's neck, def_{max}, it is necessary to introduce a tolerance factor, a coefficient $\gamma = (\varepsilon_{max} - 1)/(\varepsilon_y - 1)$, which determines the permissible *elastic* expansion of the upper edge of the cylindrical section from $\pi \cdot D_1$ to $\pi \cdot \varepsilon_{max} \cdot D_1$. The γ value is chosen between 0.8 and 0.9. The def_{max} value is determined by equating the maximum diameter of the expanded upper part of the cylindrical section, $\varepsilon_{max} \cdot D_1 = [\gamma \cdot (\varepsilon_y - 1) + 1] \cdot D_1$, and the geometrical changes that take place after the **a** → **b** transition and are expressed by Equation 6.1, that is, $[\gamma \cdot (\varepsilon_y - 1) + 1] \cdot D_1 = D_{wid}$. A solution of this equation for def_{max} gives the maximum elastic downward deflection of the bottle's neck:

$$def_{max} = h_1 - \left(h_1^2 + 0.25 \cdot (D_1 - D_2)^2 - 0.25 \cdot \left\{ \left[\gamma \cdot (\varepsilon_y - 1) + 1 \right] \cdot D_1 - D_2 \right\}^2 \right)^{0.5} \tag{6.2}$$

This model has the obvious limit for the def_{max} value; it cannot exceed the height of the transitional section h_1. The condition $def_{max} = h_1$ corresponds to the point at which the transitional section flattens (see Figure 6.1(a)). Respectively, the D_{wid} value in Equation 6.1 cannot exceed $2[h_1^2 + 0.25(D_1 - D_2)^2]^{0.5} + D_2$.

The deformation in the cylindrical part of a bottle during the **a** → **b** transition in Figure 6.1(b) is a complex process described in the science of structural mechanics as the collapse of a thin rigid shell under a distributed load. The model presented below gives the simplest semi-empirical variant of this deformation stage. Figure 6.1(c) shows the schematics of the model. The edge of the deformed cylindrical wall is stretched from $\pi \cdot D_1$ to the circumference slightly below the point of irreversible deformation, $\pi \cdot [\gamma \cdot (\varepsilon_y - 1) + 1] \cdot D_1$. Taking into account the definition of the elastic (Young's) modulus $M_{Young} = \sigma_y/(\varepsilon_y - 1)$, this upper layer is under the stress $\sigma_{top} = M_{Young} \cdot (\varepsilon_{max} - 1) = M_{Young} \cdot \gamma \cdot (\varepsilon_y - 1)$. The principal assumption of the simple top-load model is that the stress in an arbitrary elementary section of the cylindrical wall below its upper edge decreases according to the exponential law:

$$\sigma_\xi = \sigma_{top} \cdot \mathrm{Exp}(-k \cdot \xi) = M_Y \cdot \gamma \cdot (\varepsilon_y - 1) \cdot \mathrm{Exp}(-k \cdot \xi) \quad (6.3)$$

Here ξ is the distance of the section from the top of the cylinder, and k is an empirical parameter. The diameter of this section is $D_\xi = \varepsilon_\xi \cdot D_1 = (\sigma_\xi / M_{Young} + 1) \cdot D_1$. The elemental work of stretching a ring section with the cross-section $t_1 \cdot d\xi$ and the length $\pi \cdot D_1$ to the length $\pi \cdot D_\xi = \pi \cdot \varepsilon_\xi \cdot D_1$ is:

$$dW_\xi = \sigma_\xi \cdot [\text{Cross section}] \cdot [\text{Distance}] = \sigma_\xi \cdot [t_1 \cdot d\xi] \cdot [\pi \cdot (\varepsilon_\xi - 1) \cdot D_1] = \quad (6.4)$$

$$\pi \cdot \sigma_\xi^2 \cdot (t_1 \cdot D_1 / M_Y) \cdot d\xi$$

The introduction of the expression for σ_ξ from Equation 6.3 into Equation 6.4 and integration from $\xi = 0$ to $\xi = h_1$ gives the final expression for the work of deforming the cylindrical section of a bottle to the point when its elastic recovery after lifting the top load is still possible:

$$W_{\text{top-load}}^{\text{elast}} \approx (\pi \cdot D_1 \cdot t_1 / 2k) \cdot M_Y \cdot \gamma^2 \cdot (\varepsilon_y - 1)^2 \cdot [1 - \mathrm{Exp}(-2k \cdot H)] \quad (6.5)$$

Because the ε_y value does not depend much on the type of HDPE resin, Equation 6.5 provides the general assessment of the top-load test. The work of compressing the cylindrical part of a bottle to a point just before the beginning of its inelastic deformation (signifying the failure of the bottle in the test) is proportional to two parameters of the bottle: the wall thickness of its cylindrical body, t_1, and its diameter, D_1. The $W_{\text{top-load}}^{\text{elast}}$ value is also proportional to a single mechanical parameter of the resin, its stiffness (the value of the Young's modulus M_{Young}).

The maximum top load a container can carry before the start of irreversible deformation is:

$$Top\,load_{\text{elast}} = W_{\text{top-load}}^{\text{elast}} / def_{\max} \quad (6.6)$$

The maximum top load is proportional to the $W_{\text{top-load}}^{\text{elast}}$ value but it also strongly depends on the configuration of the transitional section between the (threaded) neck and the cylindrical body, as illustrated in Equation 6.2. The latter dependence explains a large variability in the design of this transitional section. The goal of the design is to achieve an even distribution of the top load across the body of the container and to avoid the development of excessive local strain areas, such as those at points P_1 and P_2 in Figure 6.1.

This simple model of the top load test provides a realistic estimation of the maximum top load. As an example, the values of $W_{\text{top-load}}^{\text{elast}}$ (Equation 6.5), $def_{\max}$ (Equation 6.2), and $Top\,load_{\text{elast}}$ (Equation 6.6) were evaluated for a small thin-walled bottle with the geometry shown in Figure 6.1 and with the following dimensions:

$D_1 = 61$ mm (2.4 in); $D_2 = 21$ mm (0.83 in); the wall thickness $t_1 = 0.35$ mm (0.014 in); the height of the transitional section $h_1 = 5$ mm (0.2 in); and the height of the cylindrical body $H = 140$ mm (5.5 in). When such a bottle is manufactured from a stiff HDPE resin characterized by a high secant 1% modulus of 150 kg/mm^2 (1,500 MPa) (the value of this modulus is close to M_{Young}) and a low yield strain ε_y of 1.02 to 1.03, the maximum reversible deformation is ~ 2 mm; the work of deformation (depending on the values of γ an k parameters) ranges from 4 to 4.5 kg·mm, and the estimated top load weight varies from 2 to 2.5 kg.

6.2 Dynamic Burst Test of HDPE Pipes and Tubing

HDPE and MDPE resins with a bimodal molecular weight distribution (see Section 1.6.4 in Chapter 1) are widely used for the manufacture of pipes (diameter > 1 cm or > 0.4 in) and tubing (diameter < 1 cm). These materials are light, corrosion-resistant, and represent the safest choice in the case of earthquakes [5]. A special type of polyethylene resin was developed for pipe manufacture and biomedical applications: radiation or chemically cross-linked polyethylene resins (PEX), which contain chemical links between macromolecules. The cross-linking makes polyethylene more resistant to creep and chemical corrosion, more durable at high temperatures, and allows the material to remain flexible under subfreezing temperatures [6–8].

Two types of tests were developed for polyethylene pipes and tubes: burst (dynamic) tests and static pressure tests [9–11]. The burst tests are carried out at temperatures from ~ 20 to 90 °C. During the burst test, the internal pressure in a pipe (air or water) is rapidly ramped to ~ 20 MPa (200 atm, 3,000 psi). Static pressure tests (sustained pressure tests) are carried out at a constant pressure below the burst pressure; they measure the time-to-failure.

The mechanical interpretation of the dynamic pressure test is straightforward; it is equivalent to the stress/strain test described in Section 1.10 of Chapter 1. As the internal pressure (the stress) increases, the pipe initially bulges at an arbitrary point (usually a defect or a bubble) where the stress in a small area of the wall exceeds the yield stress σ_y. Soon after that, the bulged area of the wall bursts under the breaking stress σ_{br}. (The σ_y and the σ_{br} values for HDPE resins are relatively close.)

Although the dynamic tests are performed by ramping the internal pressure until the pipe bursts, the meaningful task of the testing is to determine the maximum internal pressure a pipe can withstand without the wall bulging. In general, the strain at this moment can be defined as a point on the stress/strain plot just before the yield point of the material, or $\gamma \cdot (\varepsilon_y - 1) + 1$, where γ is the tolerance coefficient, 0.8 to 0.9, the same as in Equation 6.2 to Equation 6.5. The stress at this point is, respectively, $\gamma \cdot \sigma_y$. Because the yield strain (the elongation at the moment when the bulge appears) is a nearly constant value for most HDPE resins, the crucial parameter determining the outcome of the dynamic burst test (the stress at the yield point) is the resin stiffness (the Young's modulus), $M_{Young} = \sigma_y/(\varepsilon_y - 1)$ [12, 13].

6.3 Static Burst Test and Long-Term Fatigue in Polyethylene Pipes and Tubes

Static burst tests of pipes and tubes are carried out in the first range of the stress/strain curve in Figure 1.18, before the stress reaches the yield point. The behavior of many polymers, including HDPE and MDPE resins, under low stress is very different from their behavior in relatively rapid dynamic burst tests. It is governed by a phenomenon called long-term fatigue. This phenomenon in polymers (it is also called creep-to-rupture or stress cracking) is well researched.

6.3.1 Principal Equation and Theory of Low-Stress Failure

Plastic materials strained for a long time fail at a significantly lower stress than the same materials strained for a shorter time. Even a relatively small stress, 2 to 3 times lower than σ_y, if applied to a polymer article for a sufficiently long time, invariably results in its mechanical failure. In engineering practice, the dependence between the stress σ applied to a polymer sample and the time τ it takes to break the sample is usually presented in the coordinates "log(τ) as a function of σ" shown in Figure 6.2. For many polymers, including HDPE resins, such plots are linear:

$$\log(\tau) = A - K \cdot \sigma \tag{6.7}$$

Parameters A and K in Equation 6.7 depend on the mechanical properties of a resin. Equation 6.7 does not hold at very low stresses, at $\sigma \to 0$.

A general physical theory was developed to describe linear dependencies represented by Equation 6.7 as a manifestation of the universal mechanism of polymer failure [14–16]. The theory is based on numerous experimental results on the long-term fatigue of polymers at different temperatures. The most convincing results were obtained for strongly oriented polymer filaments, which usually break under stress without any noticeable elongation. When the measurement results are presented in the coordinates of Equation 6.7, they produce plots schematically shown in Figure 6.3. Such plots reflect a well-known fact that an increase in temperature accelerates the polymer failure under stress, provided that the temperature is not so high as to facilitate a substantial creep.

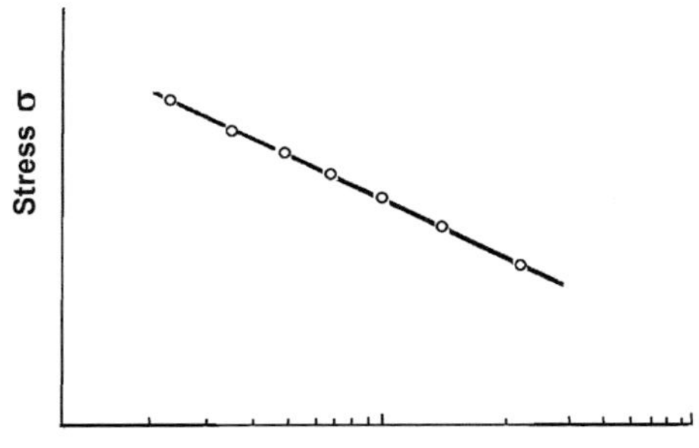

Figure 6.2
Standard presentation of long-term fatigue data for polymers

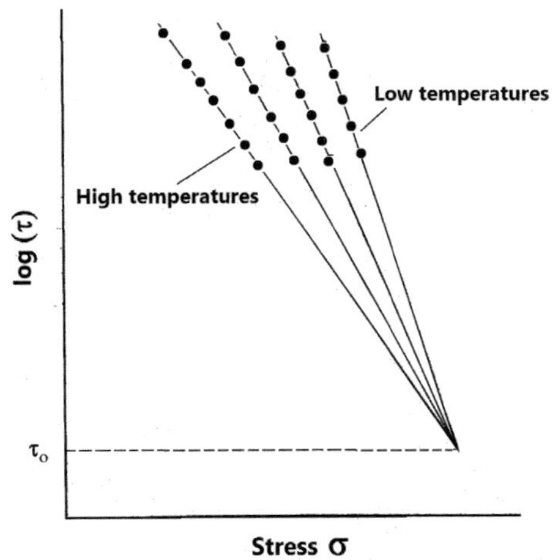

Figure 6.3
Long-term fatigue data for oriented polymer filaments in coordinates of Equation 6.7

The data on the time-to-failure of polymers, such as those in Figure 6.3, can be described by the universal expression [14]:

$$\tau = \tau_0 \cdot \text{Exp}[(U_0 - \gamma \cdot \sigma)/(k_B \cdot T)] \tag{6.8}$$

where k_B is the Boltzmann constant and temperature T is in Kelvin. The τ_0 value in Equation 6.8 is the shortest time-to-failure for any polymer. It is very small, ~ 10^{-12} to 10^{-13} sec, and it is close to the vibrational frequency of the C–C bond in a polymer chain. The U_0 term in Equation 6.8 is the initial potential energy barrier for a polymer failure. Its value, which was experimentally estimated for oriented polypropylene film, is ~ 56 kcal/mol; it is close to the activation energy of the C–C bond dissociation. The γ parameter in Equation 6.8 is the structure-dependent parameter that reflects the uniformity of stress distribution across the cross section of a polymer sample. Equation 6.8 signifies that the external stress σ applied to a polymer sample decreases the potential energy barrier for the C–C bond breaking at a particular temperature T and thus increases the frequency of a spontaneous C–C bond rupture caused by vibrational fluctuations. Thus, Equation 6.8 provides the physical meaning to parameters A and K in the empirical Equation 6.7. Detailed mechanistic studies of failure in LDPE resins at various temperatures provided significant support to this theory and supplied many quantitative details [17].

The theory represented by Equation 6.8 gives a convenient simple physical model for the description of many phenomena related to time-dependent properties of polymers, including environmental stress cracking (see Section 6.4). Experimental data for HDPE and MDPE resins with density from 0.94 to 0.95 g/cm^3 and with a high-load melt index I_{21} ~ 1 (such resins are used for the fabrication of pipes) also support this theory. When pipes made from HDPE resins are tested under the below-burst pressure, the dependence between the time-to-failure and the stress is well described by Equation 6.7 and Equation 6.8, although deviations from these dependencies are observed for other types of resins at low stresses [18, 19].

6.3.2 Physical Mechanism of Polymer Failure under Low Stress

The phenomenological mechanism of polymer failure under low stress includes several distinct stages:

craze nucleation → craze growth → craze failure (crack initiation)
→ crack propagation.

Numerous studies describing details of this cascade of events in polyethylene resins were published [8, 20–25]. Studies of the static fatigue fracture in HDPE resins identified the following features of the polymer failure under low stress [8, 23, 26]:

1. The crack initiation stage ("induction time") is the longest stage in the series of events between the stress application and the sample failure. In contrast, the crack propagation stage is, in relative terms, very rapid. Figure 6.4 gives an example of the relationship between the duration of the crack initiation and the crack propagation stages in the bending test of a notched polyethylene bar [26].
2. The rate of the slow craze-growing stage is controlled by the stress intensity [24, 25, 27, 28]. The resistance of polyethylene to the initiation of fracture under a static load is related in a large degree to its resistance to voiding [21, 26, 29]. In turn, the process of void formation depends on a combination of numerous structural factors, such as the lamella thickness, the size of amorphous regions, the degree of crystal perfection, the number of tie molecules and chain entanglements, and so on. All these factors make the prediction of resistance of polyethylene resins to crack initiation quite unreliable.

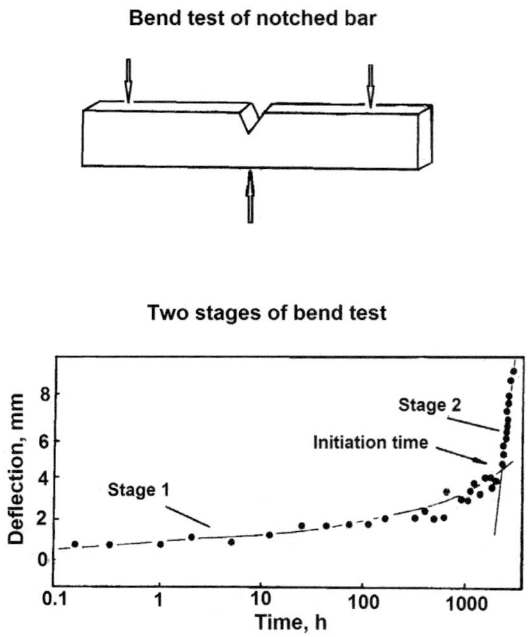

Figure 6.4
Two stages of crack failure in bend test of notched bar; data from [26]

Figure 6.5 schematically depicts the process of crack propagation through a polymer specimen [30].The scheme is based on imaging data for amorphous polymers [29] and polyethylene [21, 22] produced by the nuclear magnetic resonance technique. After the first cracks are initiated at the points of the maximum stress intensity (at the surface of a sample), the crack propagation stage (the fast stage of the fatigue failure) begins. This stage is mostly controlled by the effective viscosity of the polymer medium. Once the crazes on the path of the growing crack have nucleated, the local stress level at the tip of the crack exceeds the yield stress of the polymer. As a result, a small area between the tip of the crack and the closest craze starts to stretch and is gradually transformed into bundles of oriented macromolecules called craze fibrils. The craze fibrils steadily increase in length and eventually break. The quantitative theory of the crack propagation predicts that the crack growth rate is proportional to the yield stress of the resin and reciprocal to the square of the Young's modulus and the intrinsic viscosity of fibrils in the craze [22] (see Figure 6.5(b). Cross-linking of low-density and high-density polyethylene, either by radiation or chemical, increases the resistance to fatigue on both the crack initiation and the crack propagation stage [8, 23].

Generally speaking, the growth of oriented craze fibrils during the low-stress tests (Figure 6.5) is identical to the yielding and the neck formation in polymer samples during the standard tensile test (Section 1.10, Figure 1.18). The craze fibrils are highly oriented [17, 22, 30, 31] and they can bear a significantly greater stress than the bulk of the resin sample [26]. As obvious from Figure 6.5, the cracks propagate because the craze fibrils gradually rupture [26]. Figure 6.6 shows the end result of the slow cracking, a surface of a fractured MDPE sample covered with broken craze fibrils consisting of bundles of oriented polymer molecules [19].

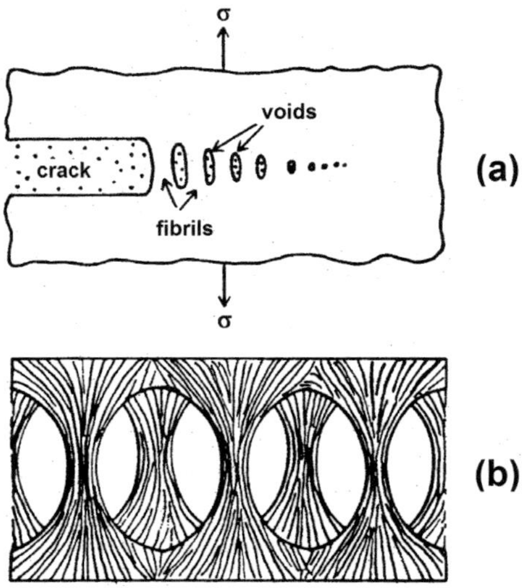

Figure 6.5
(a) Deformation of crack tip and crack propagation; (b) schematic structure of fibrils

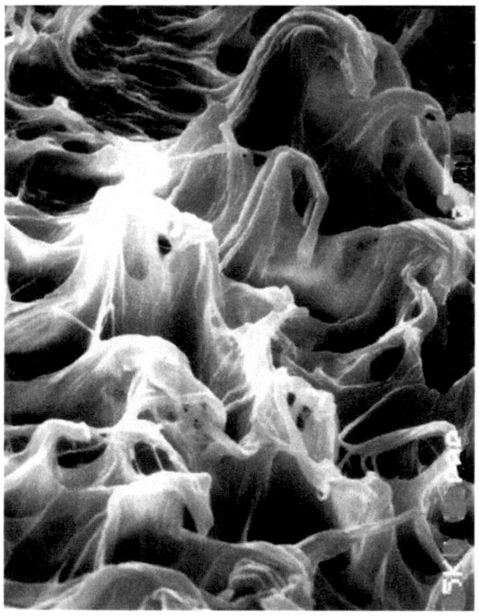

Figure 6.6
Scanning electron micrograph of fracture surface of MDPE sample; magnification ×1000

The overall mechanical behavior of HDPE and MDPE resins under low-stress conditions is predominantly affected by the macromolecular entanglement network that exists in the resin melt prior to crystallization. This entanglement network is mostly provided by macromolecules with the highest molecular weight. A similar explanation based on the entanglement mechanism is used to calculate ultimate draw ratios in polyethylene [32, 33]. The same entanglement mechanism explains why adding to a polyethylene resin a small amount of material with a high molecular weight enhances its resistance to fracture. It also explains why an increase of the average molecular weight of a polyethylene resin is always accompanied by a decrease in the crack propagation rate [34]. The same theory was also used to explain deformation and failure processes in amorphous polymers [35–37].

Polyethylene resins with a broad molecular weight distribution fail relatively easily under low stress. This deficiency is the consequence of peculiarities of polymer crystallization from the melt [38]. The crystallization process is usually accompanied by fractionation of polymer constituents. Polymer components with the highest molecular weight in such melts crystallize the first, at the highest temperature, whereas macromolecules with a low molecular weight and branched macromolecules crystallize the last, at relatively low temperatures. This fractionation can be clearly observed in the studies of polyethylene resins with the differential scanning calorimetry (DSC) method described in Chapter 3. As a consequence of the spontaneous fractionation, the low molecular weight and branched macromolecules concentrate in interlamellar regions (Figure 1.16 and Figure 1.17) and in the areas between spherulites. These macromolecules do not form good links between structural elements within spherulites (between lamellae and fibrils) and between the spherulites. These inter-spherulite regions, as well as the centers of spherulites, are the starting points of the initial crazes, which develop under stress.

The degree of interconnectedness between spherulites can be somewhat controlled by crystallization conditions. For example, rapidly quenched pipe-grade MDPE resins have relatively homogeneous morphology; their spherulites consist of lamellae that are merely ~160 Å thick. Annealing these samples induces segregation of their components into separate populations of crystallites and produces poorly connected large spherulites with thicker lamellae, ~270 Å. As a result, the annealing significantly degrades the resistance of MDPE resins to crack propagation due to a lower concentration of tie molecules [39]. In addition, a significant local stress can develop even in unstrained polyethylene articles (containers, pipes) during the later stages of crystallization. The presence of this stress also affects the crack growth rate [40].

6.4 Environmental Stress-Cracking Resistance

Resistance to environmental stress cracking (ESCR) is one of the two most important end-use properties of high molecular weight HDPE and MDPE resins formulated for applications in household and industrial containers (the second such property being stiffness). Resistance to environmental stress cracking is also an important property of LDPE and LLDPE resins used as cable and wire insulation materials.

6.4.1 Description of ESCR Test

The simplest ESCR test for polyethylene resins is called the bent strip test. It provides a good graphic description of macroscopic changes accompanying the environmental stress cracking. Figure 6.7 shows several stages of the test. A compression-molded strip of a resin is bent close to 180° and positioned in a metal holder. The bent strip is notched with a razor parallel to the long edge of the strip (to concentrate the stress in this area) and placed in a bath with detergent solution.

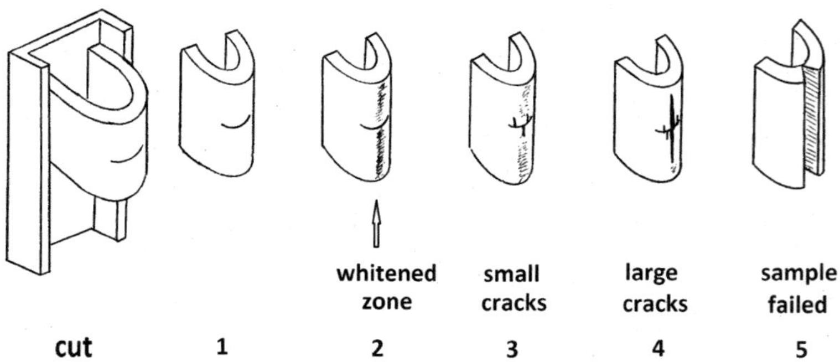

Figure 6.7 Stages of bent strip environmental stress cracking test

The most common detergents are non-ionic liquid detergents; they are usually used as 10% solutions in water. The detergents have the following chemical structure:

C_mH_{2m+1}-O-(CH_2-CH_2-O)$_n$-H with m = 12 to 18 and n = 10 to 30 or

C_9H_{19}-C_6H_4-O-(CH_2-CH_2-O)$_n$-H with n = 9 to 30.

The detergent most often used in the bent strip test is called Igepal® CO-630; it has the second of the shown formulas with n = 9.

In the beginning of the test, the sample shows no visible traces of stress for a significant period of time and the curvature of the bend remains unchanged; Stage 1 in Figure 6.7. The first manifestation of the polymer yielding to stress is an increase in the bend curvature and the development of a whitened area around the cut, indicating the formation of numerous crazes and voids in the material (Stage 2). After a certain time, one or several small cracks develop along the bend line (Stage 3). The cracks usually initiate somewhere at the original notch, which serves as a stress concentrator. Some of the cracks (those in the zone with the minimum radius of curvature and, therefore, the highest stress concentration) propagate faster than other cracks (Stage 4). In some samples, polyethylene fibrils (craze fibrils) can be seen bridging the sides of the cracks. The point when the sample fails is defined as the moment when the cracks propagate to the sides of the strip. The time-to-failure is usually determined visually by a technician who periodically examines several strips made from the same resin and submerged in the same bath with a detergent solution.

6.4.2 Physics of Environmental Stress Cracking

The original phenomenological definition of environmental stress cracking states that it is a failure in the surface-initiated brittle fracture of a polyethylene specimen under stress in contact with a sensitizing medium in the absence of which the fracture would not occur under the same stress [4, 38, 41, 42]. Although many other polymers suffer from the environmental stress cracking phenomenon [38, 41, 43, 44], this material weakness is particularly troublesome for polyethylene products made of HDPE and MDPE resins due to their wide use for the manufacture of containers for transportation and storage of household soaps and industrial detergents.

The phenomenon of environmental stress cracking is closely related to the fracture of plastics under a sub-yield stress, the phenomenon discussed in Section 6.3 [42]. Detailed studies of the rupture surface of HDPE and LDPE strips by the scanning electron microscopy method [18, 19, 45–49] showed that the fracture surfaces are always ductile. However, the size of the deformation zones, especially for HDPE resins, can be so small that the newly formed surface appears to an unaided eye as an outcome of a brittle rupture. Microscopic observations reveal however that highly oriented craze fibrils (similar to those in Figure 6.6) are indeed present on the fracture surfaces, but their size (which strongly depends on the speed of crack propagation) is usually very small.

Detergents are not the only sensitizing agents that can cause environmental stress cracking. Various alcohols, silicone oils (such as mold-release agents used during fabrication), ketones, phenols, esters, halogenated hydrocarbons, mineral and vegetable oil, and even some low-molecular-weight polymers can produce similar effects [31, 38, 47]. Environmental stress cracking is essentially a physical phenomenon; it involves neither a chemical reaction between polyethylene and a detergent nor large-scale swelling typical for solvent-initiated cracking [38]. This does not mean, however, that the surfactants do not interact with polyethylene at all. Infrared spectroscopic observations indicate that polyethylene absorbs about 1% of the detergent and water and that both compounds concentrate in amorphous regions between spherulites [50]. This localized plasticization of polyethylene is the essential feature of the environmental stress cracking phenomenon. When initial microcracks develop in a strained polyethylene sample, they propagate along the weakest boundaries in the material, which are the borders between neighboring spherulites and the borders between microfibrils and lamellae within the spherulites; that is, they propagate through the areas where the detergent/water micelles concentrate [18, 45, 51].

6.4.3 Structural Parameters of HDPE Resins Affecting ESCR

All three principal structural properties of polyethylene resins, the crystallinity degree (density), the molecular weight, and the molecular weight distribution, affect their ESCR [38].

Molecular weight: An increase in the molecular weight (a decrease in the melt index) is nearly always accompanied by the increase in the ESCR. The dependence between the time-to-failure under environmental stress cracking conditions at a particular stress level and the melt index of HDPE resins is very strong. For example, a decrease of the I_2 value of an HDPE resin from 2.0 to 0.3 results in an increase of the time-to-failure from two to over 600 hours. Similarly, a decrease of the I_2 value of an LDPE resin from 20 to 0.3 is accompanied by an increase of the time-to-failure from 20 minutes to ~7000 hours [38]. Obviously, HDPE resins with the highest possible molecular weight (from the fabrication point of view) should be used in detergent-sensitive applications.

This inverse dependence between the time-to-failure and the resin's melt index is not linear; after the melt index increases to a certain value, the ESCR decreases precipitously. The value of the melt index corresponding to the nearly complete loss of the ESCR is different for different resin types and should be experimentally determined in each particular case. In addition, the dependence between the ESCR and the melt index is affected by variations in other properties of the resins. As Figure 6.8 demonstrates, these effects make the correlation between the ESCR and the melt index for HDPE resins in the density range of 0.950 to 0.955 g/cm^3 less clear-cut [46].

Molecular weight distribution: The width of the molecular weight distribution significantly affects the ESCR [38]. Polymer components with a low molecular weight, which are always present in resins with a broad molecular weight distribution, are especially detrimental for the ESCR [38]. Extraction of these components leads to a very large increase in the ESCR, and concurrently, addition of low molecular weight waxes to polyethylene resins decreases their ESCR [38]. The ESCR is the highest for HDPE resins with a large fraction of high molecular weight macromolecules [18, 52]. The longest polymer chains in these resins are the source of interlamellar ties, which play the crucial role in the ESCR [53].

A direct quantitative assessment of the concentration of tie molecules has proved to be difficult, but several indirect experimental techniques have been developed. They include polarization IR spectroscopy [54], crystallization fractionation [55], and step-wise isothermal crystallization [13]. One such technique involves deformation of MDPE or LLDPE film followed by a chlorination reaction and by analysis of polarized IR spectra. This technique provides the means of ranking resins of comparable density but with a different molecular weight and molecular weight distribution in terms of their resistance to crack propagation [54]. The tie molecules can be also partially separated from the bulk of HDPE resins through the use of the preparative crystallization fractionation (Crystaf) technique with 1,2,4-trichlorobenzene as a solvent. The polymer fraction rich in tie molecules crystallizes under these conditions at temperatures between 75 and 85 °C [55]. The ESCR of the resins increases when the molecular weight and the content of this fraction increase.

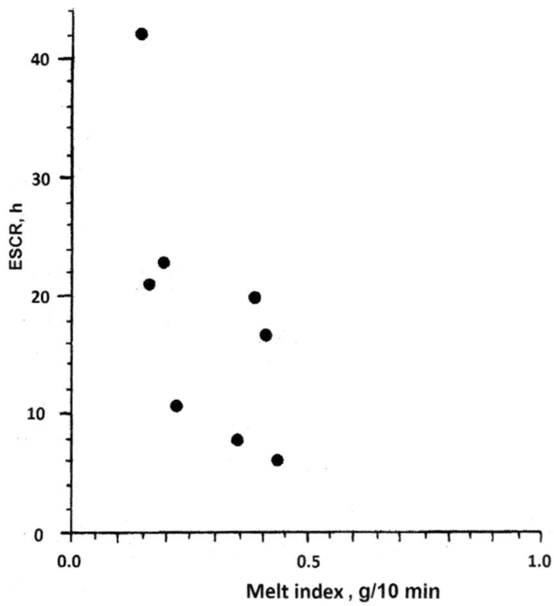

Figure 6.8
Effect of melt index on ESCR of HDPE resins; data from [46]

Crystallinity level: The effect of the crystallinity degree (density) on the ESCR is also very difficult to present in a concise form. Depending on the testing conditions, an increase in the crystallinity degree can produce either an increase or a decrease in the ESCR [38]. The bent strip ESCR test shown in Figure 6.7 is essentially a constant strain test. Therefore, polymer components with a low crystallinity degree in the strips are subjected to a lower stress level at the same strain compared to the highly crystalline components, since the tensile modulus of polyethylene resins of low crystallinity (MDPE, LLDPE, LDPE) is lower (Section 1.10). Consequently, a reduction in the crystallinity degree and density generally leads to a higher ESCR [38, 46]. However, experimental data relating the ESCR and the polyethylene density are usually very scattered and are affected by other factors. The most important of them are the variations in the molecular weight and the molecular weight distribution (Figure 6.9), as well as the length of side-groups in ethylene/α-olefin copolymers [56]; the longer the side chain (for example, in ethylene/1-octene copolymers) the higher is the ESCR.

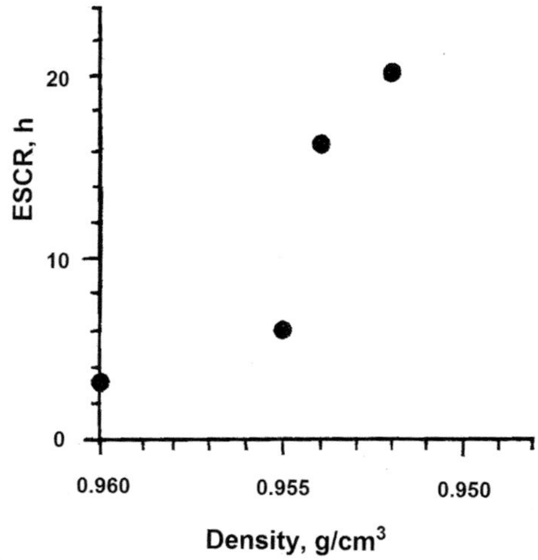

Figure 6.9
Effect of density on ESCR of HDPE resins; data from [46]

In addition to structural parameters of polyethylene resins, their ESCR also depends on numerous processing parameters [18, 38, 57]: the melt temperature, the speed of extrusion, the speed of cooling, melt orientation during fabrication, and other factors. These parameters affect morphological features of the articles, such as the size and the quality of spherulites and crystallites, chain orientation, the degree of chain entanglement, and so on. None of these effects have yet been studied in sufficient detail to yield a simple and coherent set of rules. However, the existence of these effects imposes significant restrictions on the interpretation of ESCR test results.

6.4.4 Relationship between ESCR and Long-Term Fatigue in Polyethylene

On the quantitative level, general features of environmental stress cracking can be better understood when this type of polymer failure is analyzed within the framework of the long-term stress fatigue (Section 6.3.1 and Section 6.3.2). The theory that describes the linear dependencies shown in Equation 6.7 proposes the general polymer failure mechanism [14, 15] represented by the universal expression shown in Equation 6.8. This theory is also a convenient starting point for the description of many phenomena related to environmental stress cracking. Both types of failure were studied in parallel by subjecting pipes manufactured from high molecular weight HDPE and MDPE resins ($I_{21} \sim 1$, density from 0.94 to 0.95 g/cm^3) to internal pressure [18, 19]. In the absence of a surfactant, the dependence between the time-to-failure and the stress for HDPE samples is well described by Equation 6.7. However, when the testing is carried out in the presence of a detergent (1% of Igepal® CO-630 solution in water), a clear deviation from the linear logarithmic dependence is observed in the high τ/low σ range. Figure 6.10 and Figure 6.11 emphasize the surfactant effect by plotting the experimental data from ref. [18] in the coordinates "σ_d/σ_a as a function of log(τ)," where σ_d and σ_a are the stresses in the presence of the detergent solution and in air, respectively. Thus, the σ_d/σ_a ratio is the measure of the environment-related weakening of polyethylene.

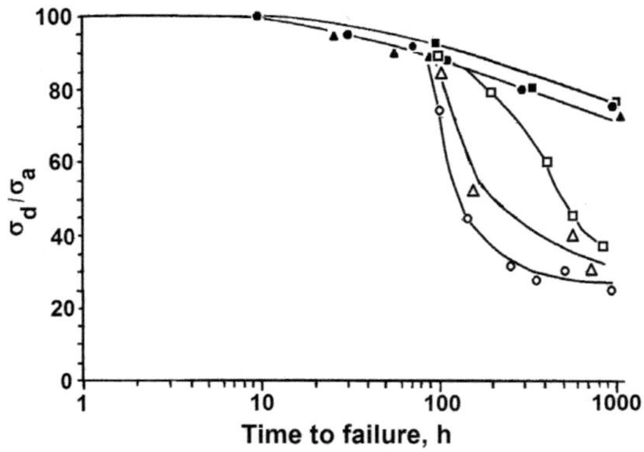

Figure 6.10
Effect of detergent solution on long-term fatigue of HDPE resins: ▲, ■, ● - HDPE resins with high ESCR; △, □, o - HDPE resins with low ESCR. Experimental data from [18]

Plotting experimental data in these coordinates allows a separation of two effects: the universal time-dependent polymer properties discussed in Section 6.3 and the environmental stress cracking effect as such. Plotting in these coordinates also provides the answer to an important question about the environmental stress cracking: what reduction in stress is required to enable an HDPE or a MDPE sample to survive under some particular conditions for a particular period of time?

The plots in Figure 6.10 and Figure 6.11 identify two types of environmental stress cracking behavior. Some HDPE samples in Figure 6.10 have a good ESCR; their rupture stress is reduced by only ~ 20% after 1,000 hours of testing in the presence of the detergent (top curves). Other HDPE samples have a very poor ESCR; to survive for 1,000 hours, their stress should not be higher than 25 to 30% of the stress they can bear in air (bottom curves). The comparison of different curves in Figure 6.10 shows also that the HPDE samples that have longer lifetimes at low stresses also have a better resistance to the environmental stress cracking: cracking-resistant samples survive for a much longer time (both in air and in the presence of a detergent) than cracking-prone samples.

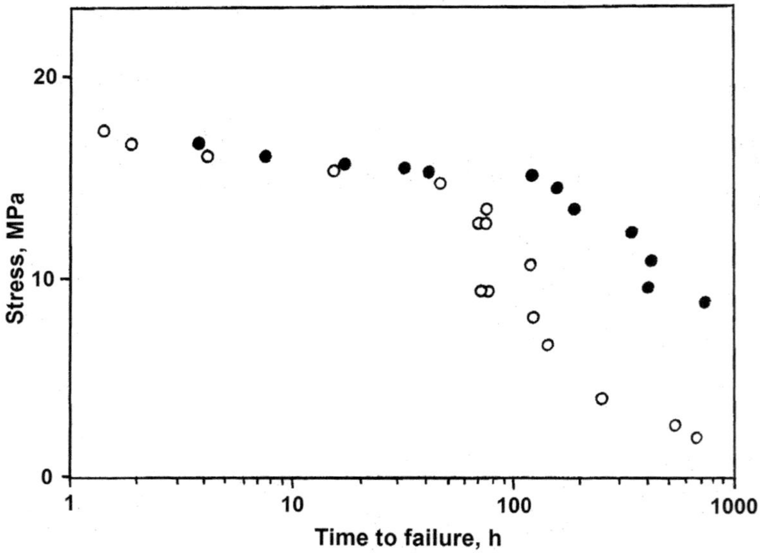

Figure 6.11 Correlation between stress and time-to-failure for two MDPE copolymers (ethylene/1-butene with density 0.940 g/cm^3); data from [19]

Several features related to the loss of strength under environmental stress cracking conditions for HDPE resins follow from the analysis of the data in Figure 6.10 and Figure 6.11:

1. The resistance of polyethylene to environmental stress cracking at a relatively high stress (close to the yield stress) does not decrease appreciably in the presence of a detergent, most probably, because the survival time under these conditions, a few hours, is not long enough for the environment-related cracks to develop. The environmental stress cracking can be clearly observed only at a low stress level when the time-to-failure exceeds 50 hours.
2. The resistance to environmental stress cracking correlates with long-term fatigue (creep) properties, which suggests a similarity in the rupture mechanisms [16]. The changes in the polyethylene structure or processing parameters that improve its long-term stability also improve its ESCR.
3. The fact that a significant period of time is required for the environmental stress cracking to manifest itself suggests that it is a diffusion-related process. Either a significant time is required for detergent solution to diffuse into a resin or the detergent-induced structural changes involve rearrangements of polymer chains which are controlled by diffusion.

These features of environmental stress cracking suggest a modified definition of the environmental stress cracking phenomenon [18, 58]: environmental stress cracking is the acceleration of the long-term stress-fatigue process in polyethylene caused by several types of surface-active agents that facilitate crack propagation.

6.4.5 Mechanism of Environmental Stress Cracking

The mechanism of the polyethylene failure under the environmental stress cracking conditions represents a special case of the general mechanism describing the mechanical failure of polymers under low stress, which is discussed in Section 6.3.2. Similar to the process of failure under low stress, the environmental stress cracking mechanism includes four distinct stages [59, 60]:

craze nucleation (the longest stage) → craze growth → craze failure (crack initiation) → crack propagation.

The overall behavior of polyethylene articles under the conditions of environmental stress cracking is similar to that of their failure under low stress. Both processes are dominated by the molecular entanglement network, which is formed in the polymer melt prior to crystallization [38] (Section 6.3.2). Crystallization of most polyethylene resins from the melt leads to a spontaneous fractionation of their components. The low molecular weight and branched macromolecules crystallize at the latest stage of crystallization and concentrate in interlamellar and inter-spherulite regions. These short polymer chains cannot serve as good ties between lamellae and between spherulites. The principal feature responsible for the ESCR is the presence of tie molecules of a high molecular weight, which reinforce mechanically weak areas between and within spherulites and mitigate the effects of polymer fractionation [18, 38, 45–49]. The content of tie molecules in an HDPE resin can be artificially increased, for example, by co-blending a small amount of a semi-crystalline ethylene block copolymer with the resin [61] or by co-blending an HDPE resin with a high molecular weight LLDPE resin, especially if the LLDPE resin contains a small amount of long-chain branches [62].

Two subjects are important for the discussion of the mechanism of environmental stress cracking: the role of the detergent and the role of the solvent for the detergent. A detailed analysis of the liquid environment and its effect on the ESCR of LDPE and HDPE resins showed several important effects [17, 27, 30, 41, 52, 60].

Figure 6.12 compares schematically the crack propagation phenomena in an HDPE resin in three different environments: air, water, and 10% water solution of Igepal CO-630 [27]. The dependence between the stress and the speed of crack propagation measured in the air test shows the complex fatigue behavior typical for HDPE resins in general (Section 6.3.2). The cracks start to grow at a certain finite stress level. Initially, even a small increase of the stress leads to a rapid increase of the crack propagation rate but the acceleration of the crack propagation becomes less pronounced at high stresses. The behavior of the same HDPE resin in the water environment is mostly the same as in air, that is, water is not an effective sensitizing agent. An introduction of the detergent changes several features of the dependence between the stress and the crack growth rate:

1. The minimum stress required for the crack to start growing decreases about twofold.
2. A new stage in the crack propagation process becomes apparent: when the stress reaches a certain level, the speed of the crack growth does not depend on the stress.
3. The environmental stress cracking phenomena manifest themselves at relatively low stresses. The specific effect of a detergent disappears in the high-stress range of the curve in Figure 6.12 but this range (the range of a very short time-to-failure) is of little concern under the real-life conditions.
4. A very large difference exists in the speed of crack propagation between solutions of Igepal CO-630 in different solvents (Figure 6.13).

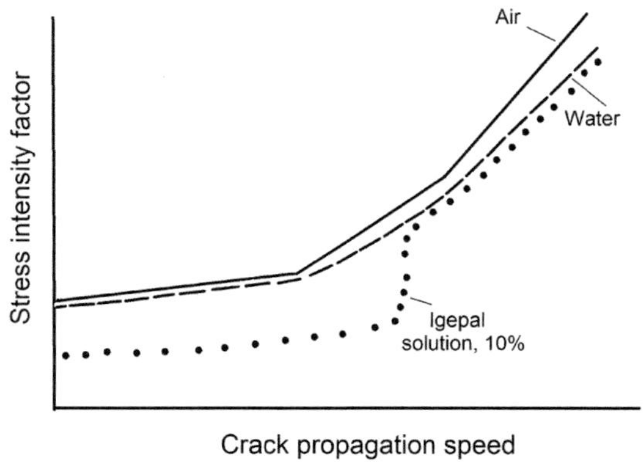

Figure 6.12
Crack propagation speed in HDPE sample in three environments; data from [30]

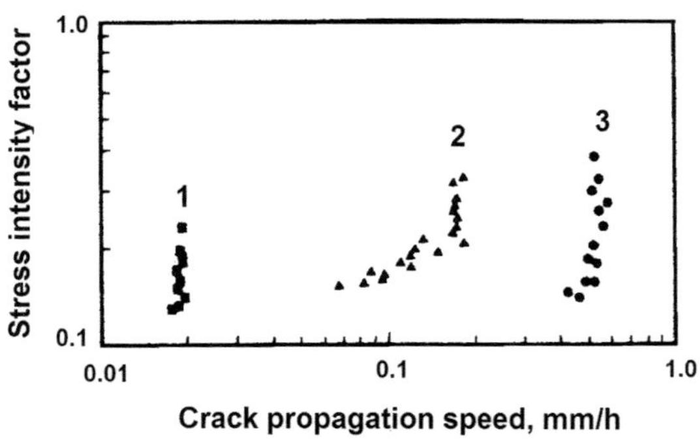

Figure 6.13
Environment effect on crack propagation speed in HDPE resin; data from [30]. Solvents for Igepal CO-630: (1) ethanol, (2) ethylene glycol, and (3) water

The ESCR of polyethylene resins strongly depends on the concentration of a detergent in water (Igepal CO-630 in this case) [63, 64]. The largest reduction in the time-to-failure occurs at a low detergent concentration of ≤0.1%. An additional increase of the detergent concentration up to ~20 to 25% leads to a relatively small incremental reduction of the time-to-failure; the effect is minimal when the detergent concentration reaches 50%, and a further increase of the detergent concentration from 50 to 100% does not produce any effect at all.

Liquid organic compounds of low molecular weight, when used by themselves, also exhibit some environmental effect. This effect is independent of such properties of the liquids as viscosity and surface tension. The environmental effect of detergent/solvent combinations strongly depends on the interaction between the components of the mixtures. Cracks in HDPE resins propagate relatively slowly when the resins are exposed to 10% solution of Igepal CO-630 in ethanol, whereas colloidal solutions of the detergent in ethylene glycol or water are 20 to 30 times more aggressive. The effect of Igepal CO-630 solution in ethylene glycol increases with the detergent concentration and becomes more pronounced after the critical micelle concentration is reached. Increasing the detergent concentration from 0.001 to 2% produces a 25-fold increase in the crack growth rate [17].

The environmental stress cracking effect of detergent solutions strongly depends on temperature [17, 51, 57]. This dependence can be described by the Arrhenius equation. The activation energy of the process does not depend on the detergent concentration but it strongly depends on the hydrophilic-lipophilic balance of the detergent. Detergents usually used for the ESCR testing exhibit a combination of both lipophilic properties due to their oxyethylene chains and hydrophilic properties due to the alcohol groups, the end-groups of the detergent molecules. An increase in the lipophilic activity (an increase in the length of the oxyethylene chain) produces an increase in the activation enthalpy of crack propagation.

The application of hydrostatic pressure to the detergent solution reduces the crack propagation speed [17]. This effect can be apparently explained by the fact that the crazing process is accompanied by an increase in the volume of the sample and therefore would be hindered as the hydrostatic pressure increases. When silicone oils are used in environmental stress cracking tests, an increase in oil viscosity is accompanied by a noticeable decrease in the crack propagation speed [27].

Based on all these observations, a plasticization model of the environmental stress cracking phenomenon was proposed [30, 31, 60]. The model is based on the general crack propagation model shown schematically in Figure 6.5. Thermodynamic calculations indicate that a significant stress in two areas of a strained polymer sample, at the tip of the crack and in the craze that precedes the crack tip, both greatly increase the permeability of the material. This change in polymer permeability produces a local swelling, a significant plasticization of the amorphous phase, and the reorganization of the chain entanglement. All these effects lead to a decrease in the load-bearing capability of the amorphous phase. Indeed, electron microscopic studies confirmed the effect of aggressive agents on the polymer morphology at the deformation stage [46, 47, 51, 60]. The chemicals that cause environmental stress cracking decrease cohesion between lamellae and microfibrils in spherulites and thus facilitate the crack propagation. Although the environmental stress cracking is inherently a slow-developing process, it is nevertheless affected by some agents with low viscosity. It is possible that the particularly aggressive properties of detergent solutions in water can be partially attributed to their lower effective viscosity. Although this mechanism of environmental stress cracking is well supported by the experimental evidence, it is the least persuasive argument in explaining the very high effectiveness of colloidal detergent solutions as agents of environmental stress cracking compared to other organic liquids, which also swell polyethylene.

References

1. Standard Test Method for Column Crush Properties of Blown Thermoplastic Containers, (ASTM) D2659-11; *http://www.astm.org/Standards/D2659.htm*
2. Standard Test Method for Compression Resistance of a Container under Constant Load, ASTM D4577-05 (2010); *http://www.astm.org/Standards/D4577.htm*
3. Column Crush Properties of Blown Thermoplastic Containers, ASTM D2659-95; *http://www.astm.org/DATABASE.CART/HISTORICAL/D2659-95.htm*
4. Deutsches Institut für Normung E. V. (DIN) 55440-1:99; Packaging Compression Test
5. Gahleitner, M., *eXPRESS Polym. Lett.* (2011) 5, p. 936, doi:10.3144/expresspolymlett.2011.91
6. Standard Specification for Crosslinked Polyethylene (PEX) Tubing, ASTM F876-10e1; *http://www.astm.org/Standards/F876.htm*
7. Standard Specification for Crosslinked Polyethylene (PEX) Hot- and Cold-Water Distribution Systems, ASTM F877-11a; *http://www.astm.org/Standards/F877.htm*
8. Baker, D.A., Bellare, A., Pruitt, L., *J. Biomed. Mater. Res., Part A* (2003) 66, p. 146
9. Test Method for Time-to-Failure of Plastic Pipe Under Constant Internal Pressure, ASTM D1598-02 (2009); *http://www.astm.org/Standards/D1598.htm*
10. Test Method for Resistance to Short-Time Hydraulic Pressure of Plastic Pipe, Tubing, and Fittings, ASTM D1599-99 (2011); *http://www.astm.org/Standards/D1599.htm*
11. Standard Test Method for Obtaining Hydrostatic Design Basis for Thermoplastic Pipe Materials or Pressure Design Basis for Thermoplastic Pipe Products, ASTM D2837-11; *http://www.astm.org/Standards/D2837.htm*
12. Bohm, L.L., Enderle, H.F., Fleifßner, M., *Adv. Mater. Sci.* (1992) 4, p. 234
13. Cazenave, J., Seguela, R., Sixou, B., Germain, Y., *Polymer* (2006) 47, p. 3904
14. Zhurkov, S.N., *J. Polym. Sci., Part A: Polym. Chem.* (1972) 10, p. 1509
15. Kausch, H.H., Williams, J.G., In *Encyclopedia of Polymer Science and Engineering*, Mark, H.F., Kroschwitz, J.I. (Eds), vol. 7 (1986) John Wiley & Sons, New York, p. 360
16. Rose, L.J., Channell, A.D., Frye, C.J., Capaccio, G., *J. Appl. Polym. Sci.* (1994) 54, p. 2119
17. Ohde, Y., Okamoto, H., *J. Mater. Sci.* (1980) 15, p. 1539
18. Lustiger, A., Markham, R.L., *Polymer* (1983) 24, p. 1647
19. Lustiger, A., Corneliussen, R.D., *J. Mater. Sci.* (1987) 22, p. 2470
20. Friedrich, K., *Adv. Polym. Sci.* (1983) 52–53, p. 225
21. Kasakevich, M.L., Moet, A., Chudnovsky, A., *J. Appl. Polym. Sci.* (1990) 39, p. 395
22. Brown, N., Lu, X., *Polymer* (1985) 36, p. 543
23. Pruitt, L.A., *Biomaterials* (2005) 26, p. 902
24. Parsons, M., Stepanov, E.V., Hiltner, A., Baer, E., *J. Mater. Sci.* (2000) 35, p. 2659

25. Favier, V., Giroud, T., Strijko, E., Hiver, J.M., G'Sell, C., Hellinckx, S., Goldberg, A., *Polymer* (2002) 43, p. 1375
26. Barry, D., Delatycki, O., *J. Polym. Sci, Polym. Phys. Ed.* (1987) 25, p. 883
27. Tonyali, K., Rogers, C.E., Brown, H.R., *Polym. Eng. Sci.* (1987) 27, p. 82
28. Tonyali, K., Rogers, C.E., Brown, H.R., *J. Macromol. Sci., Part B: Phys.* (1989) 28, p. 235
29. Adriaensens, P., Storme, L., Carleer, R., Vanderzande, D., Gelan, J., Litvinov, V.M., Marissen, R., *Macromolecules* (2000) 33, p. 4836
30. Tonyali, K., Rogers, C.E., Brown, H.R., *Polymer* (1987) 28, p. 1472
31. Brown, H.R., *Polymer* (1986) 19, p. 1186
32. Smith, P., Lemstra, P.J., Booij, H.C., *J. Polym. Sci, Part B: Polym. Phys.* (1981) 19, p. 877
33. Irvine, P.A., Smith, P., *Macromolecules* (1986) 19, p. 240
34. Marshall, G.P., Culver, L.E., Williams, J.G., In *Plastics and Polymers* (1970) 38, p. 95
35. Donald, A.M., Kramer, E.J., *Polymer* (1983) 24, p. 1063
36. Donald, A.M., Kramer, E.J., *Polymer* (1982) 23, p. 461
37. Kramer, E.J., In *Developments in Polymer Fracture-1*, Andrews, E.H. (Ed.) (1979) Applied Science Publishers, London, Chapter 3
38. Howard, J.B., In *Engineering Design for Plastics*, Baer, E. (Ed.) (1964) Reinhold Publishing Corp., New York, p. 742
39. Strebel, J.J., Moet, A., *J. Polym. Sci, Part B: Polym. Phys.* (1995) 33, p. 1969
40. Lu, X., Brown, N., *Polymer* (1987) 28, p. 1505
41. Wright, D.C., *Environmental Stress Cracking of Plastics* (1996) Rapra Technology, Shawbury, Great Britain, Chapter 1
42. Ayyer, R., Hiltner, A., Baer E., *J. Mater. Sci.* (2007) 42, p. 7004
43. Williams, J.G., Marshall, G.P., *Proc. R. Soc. A* (1975) 342, p. 55
44. Howard, J.B., In *Crystalline Olefin Polymers, Part II*, Raff, R.A.V., Doak, K.W. (Eds.) (1964) Interscience, New York
45. Hase, T.W., McRae, P.H., *SPE J.* (1968) 24, p. 27
46. Hannon, M.J., *J. Appl. Polym. Sci.* (1974) 18, p. 3761
47. Singleton, C.J., Roche, E., Geil, P.H., *J. Appl. Polym. Sci.* (1977) 21, p. 2319
48. Bandyopadhyay, S., Brown, H.R., *Polymer* (1978) 19, p. 589
49. Bandyopadhyay, S., Brown, H.R., *J. Mater. Sci.* (1977) 12, p. 2131
50. Frayer, P.D., Tong, P.P., Dreher, W.W., *Polym. Eng. Sci.* (1977) 17, p. 27
51. Ghanbari-Siahkali, A., Kingshott, P., Breiby, D.W., Arleth, L., Koch, C., Almdal, K.K., *Polym. Degrad. Stab.* (2005) 89, p. 442
52. Aiba, M., Osawa, Z., *Polym. Degr. Stab.* (1998) 61, p. 1
53. Lustiger, A., Corneliussen, R.D., *J. Polym. Sci, Part B: Polym. Phys.* (1986) 25, p. 1625
54. Lustiger, A., Ishikawa, N., *J. Polym. Sci, Part B: Polym. Phys.* (1991) 29, p. 1047
55. Soares, J.B.P., Abbott, R.F., Kim, J.D., *J. Polym. Sci, Part B: Polym. Phys.* (2000) 38, p. 1267
56. Bubeck, R.A., Baker, H.M., *Polymer* (1982) 23, p. 1680
57. Soni, P.L., Geil, P.H., *J. Appl. Polym. Sci.* (1979) 23, p. 1167
58. Lustiger, A., In *Failure of Plastics* (1986) Hanser, Munich, p. 305
59. Bubeck, R.A., *Polymer* (1981) 22, p. 682
60. Lagaron, J.M., Pastor, J.M., Kip, B.J., *Polymer* (1999) 40, p. 1629

61. Lustiger, A., Avinash, G., *U. S. Patent* 6 153 703 (2000)
62. Schellenberg, J., Fienhold, G., *Polym. Eng. Sci.* (1998) 38, p. 1413
63. Qian, R., Lu, X., Brown, N., *Polymer* (1993) 34, p. 4727
64. Tonyali, K., Brown, H. R., *J. Mater. Sci.* (1987) 22, p. 3287

7 End-Use Testing of LLDPE Film; Extractables, Solubles, Blocking, and Haze

7.1 Introduction

The four tests discussed in this chapter, the measurement of amounts of extractable and soluble material in LLDPE film, blocking of LLDPE film, and haze of polyethylene film, although at the first glance they appear to measure completely unrelated properties of polyethylene products, all reflect one of the most basic features of catalysts used for the synthesis of ethylene/α-olefin copolymers.

Currently, most of the catalysts used for the synthesis of ethylene/α-olefin copolymers with a low content of α-olefin are either heterogeneous (particulate solids) or they form solid products when solutions of the catalyst components are mixed in polymerization reactors - see Section 1.2 in Chapter 1. All major types of these catalysts, including supported Ziegler-Natta catalysts, titanium-based catalysts used in solution polymerization processes, and chromium oxide catalysts, contain several types of active centers. The centers differ in three respects [1]:

1. Different active centers produce polymer molecules of a different average molecular weight. Ethylene copolymers produced with all these catalysts have a broad molecular weight distribution. This effect is described in Section 1.5 and Section 1.6 in Chapter 1. The main manifestation of this variability in molecular weight, high melt flow ratios of such resins, is described in Chapter 2.
2. Different active centers have a different ability to copolymerize ethylene with α-olefins. This feature of the active centers is described in Section 1.8 in Chapter 1. As a result, different copolymer molecules in the products have a different composition. Some active centers exhibit very poor copolymerization ability. These centers produce copolymer molecules with a very low α-olefin content. Other active centers copolymerize α-olefins with ethylene quite well and produce copolymer molecules with a high α-olefin content. In contrast, LLDPE resins produced with metallocene and some post-metallocene catalysts are usually compositionally uniform. When these resins are solvent-fractionated, the fractions may differ in molecular weight but their α-olefin content is approximately

the same. Section 1.8 describes experimental techniques for the evaluation of the compositional nonuniformity of LLDPE resins and gives several examples of resins produced with Ziegler-Natta and chromium oxide catalysts.

3. Different active centers have different kinetic stability. The centers that do not copolymerize α-olefins with ethylene well are more stable whereas the centers that copolymerize α-olefins with ethylene well are usually quite unstable. As a consequence, the ratio between the yields of copolymer molecules with the low and the high α-olefin content can vary depending on the residence time of catalyst particles in a polymerization reactor.

Numerous experimental observations showed that the two properties of the active centers, their ability to form copolymer molecules of a particular molecular weight and the ability to produce copolymers of a particular composition, parallel each other: the higher the molecular weight, the lower is the α-olefin content in the copolymer molecules. Table 7.1 gives one example of this trend in the case of an ethylene/1-hexene copolymer produced with a commercial supported Ziegler-Natta catalyst. The detailed experimental analysis showed [1] that the catalyst has five types of active centers (they are marked Centers I to V), which differ both in the average molecular weight of the macromolecules they produce and in their ability to copolymerize α-olefins with ethylene. The table compares the weight-average molecular weight and the composition of copolymer fractions produced by different types of active centers. The average 1-hexene content in the total copolymer (C_M^{copol}) is ~3 mol %, a typical composition of an LLDPE resin, but, in reality, this resin consists of five copolymer components with very different molecular weight and composition. The components range from a completely amorphous elastomeric material of low molecular weight (C_M^{copol} ~25 mol %, M_w ~9,000) to a highly crystalline, high molecular weight material barely distinguishable from an ethylene homopolymer (C_M^{copol} ~0.2 mol %, M_w ~320,000).

Table 7.1 Characteristics of Active Centers in Commercial Ziegler-Natta Catalyst [1]

Active Center:	I	II	III	IV	V
Weight-average molecular weight M_w:	9,200	23,200	51,400	124,000	316,500
C_M^{copol}, mol %:	~25	2.8–3.0	~0.8	~0.3	~0.2

All the end-use properties of LLDPE film discussed in this chapter are the consequences of this nonuniformity of copolymer components in terms of their α-olefin content and molecular weight.

The detailed composition analysis of copolymer components produced by different types of active centers, such as the data in Table 7.1, is a complex and a time-consuming process. In everyday practice, a much simpler analysis is usually performed, fractionation of LLDPE resins with different solvents. Table 7.2 gives one example of such fractionation for an LLDPE resin, an ethylene/1-hexene copolymer produced with a commercial Ziegler-Natta catalyst [2]. The average content of 1-hexene in the resin is 3.6 mol %. A comparison of Table 7.1 and Table 7.2 shows that extraction of the resin with hexane at room temperature separates the copolymer material mostly produced by active centers I in the catalyst and the additional extraction with boiling hexane removes a part of the material produced by centers II.

Table 7.2 Solvent Fractionation of LLDPE Resin (C_M^{copol} = 3.6 mol %, density 0.917 g/cm^3) Produced with Commercial Ziegler-Natta Catalyst [2]

Solvent	Temperature, °C	Fraction, %	C_M^{copol}, mol %
Hexane (cold)	20	7.6	~25
Hexane (boiling)	68	4.0	9.6
Heptane (boiling)	98	4.6	9.1
Octane (boiling)	126	12.2	5.4
Residue		71.5	1.2

7.2 Extractables and Solubles in LLDPE Resins

7.2.1 Measurement Methods

The requirement for determining hexane extractables in LLDPE resins is prescribed by the U. S. Code of Federal Regulations Title 21 (21 CFR 177.1520), which limits the amount of extractable polymer material in food-packaging containers and in film items. The test is very simple (see several examples in refs. [3–5]). It is performed on 4-mil (0.004 in, ~100 micron)-thick LLDPE film, either cast or blown. A 2.5 g piece of film is cut into square-inch sections and placed into a cylindrical stainless-steel wire basket. The basket is placed into a 2-liter round-bottom glass vessel half-filled with hexane and heated to 50 °C. After a two-hour extraction, the basket is removed from the vessel, the film is rinsed with hexane and dried at 80 °C under vacuum for two hours. The extractables content is calculated from the weight loss of the film.

From the polymer chemist's standpoint, as exemplified by the data in Table 7.1 and Table 7.2, extraction with hexane at 50 °C, i.e., half-way between room temperature and the hexane boiling point, completely removes from a sample all the highly branched amorphous material produced by centers I and a part of the material produced by centers II.

An alternative technique for the measurement of highly branched amorphous components in LLDPE resins is based on the measurement of the parameter called "xylene solubles". This parameter is measured by the following procedure. Two grams of an LLDPE resin (in any physical form, powder, pellets, or film) is placed in a 500-ml round-bottom flask containing 200 ml of *ortho*-xylene. The solvent is refluxed under stirring until the sample is completely dissolved, and then the flask is cooled in a water bath to 25 °C. After most of the resin crystallizes, the precipitate is separated by filtering and dried. The content of xylene solubles is calculated by comparing the weight of the dried precipitate and the weight of the original LLDPE sample.

The amount of xylene solubles in LLDPE resins is usually significantly higher than the amount of hexane extractables. For example, ethylene/1-butene LLDPE resins with density 0.919 g/cm^3 and the I_2 melt index of 0.9–1.0 g/10 min contain 0.7 to 1.4% of hexane extractables but from 7 to 10% of xylene solubles [4]. The xylene dissolution/precipitation procedure removes from the resins both the highly branched amorphous material produced by active centers I and all the moderately branched material produced by centers II.

Both techniques for the measurement of highly branched components in LLDPE resins suffer from several shortcomings: they are tedious, labor-intensive and employ large quantities of flammable organic solvents. Several attempts were undertaken to replace them with automatic procedures. One such technique is the crystallization fractionation (Crystaf) method described in Section 1.8 of Chapter 1. As far as the measurement of amorphous components in LLDPE resins is concerned, this technique can be viewed as an expanded and automated version of the xylene solubles method. In the first step in the Crystaf method, a small resin sample is dissolved at a high temperature (130 to 140 °C) in an organic solvent, *ortho*-dichlorobenzene or 1,2,4-trichlorobenzene. The hot solution is very slowly cooled resulting in a gradual crystallization of copolymer fractions. The cooling continues down to 30 °C. At this point, all crystallizable material is precipitated and the solution contains only the amorphous material. As an example, Figure 7.1 shows the Crystaf crystallization curve of an LLDPE resin, the ethylene/1-hexene copolymer with C_M^{copol} = 2.6 mol % (density 0.920 g/cm^3). The curve is resolved into curves of several components of different composition, which crystallize at different temperatures [6]. This particular resin has a broad compositional distribution. Two of its components are highly crystalline materials with a low 1-hexene content, one component with C_M^{copol} ~ 0.4 mol % (the peak at ~ 82 °C), which is mostly produced by active centers V (see Table 7.1). Another highly crystalline component with C_M^{copol} ~ 0.6 mol % (the peak at ~ 79 °C) is mostly produced by centers IV. The resin also has several components of a low crystallinity degree. Their crystallization temperatures vary from 71 to 42 °C. These fractions are produced by active centers II and III (such materials can partially co-crystallize) and their C_M^{copol} values vary from ~1 to >5 mol %. In addition, this resin contains ~11% of a completely amorphous material which remains in solution at 30 °C, practically the same LLDPE component as the xylene-soluble fraction. This copolymer component is mostly the material produced by centers I.

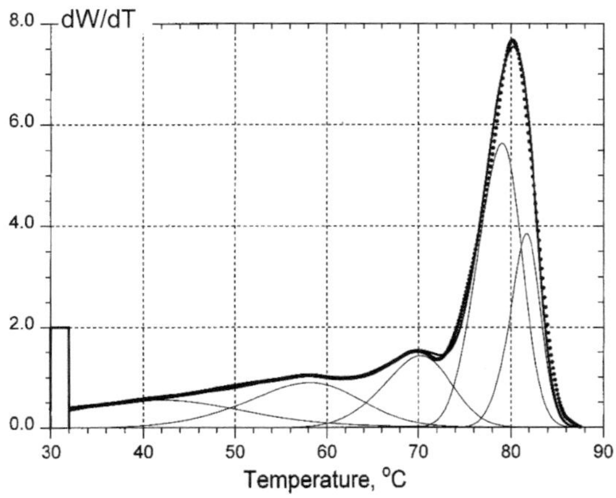

Figure 7.1
Crystaf curve of LLDPE resin with C_M^{copol} = 2.6 mol % (points) and its resolution into peaks of individual components

An automated technique for the measurement of hexane extractables in LLDPE film was also developed [7]. The method uses the FIPA technique (flow injection polymer analysis), a fast and accurate procedure of a simultaneous measurement of the molecular weight, the intrinsic viscosity, and the hydrodynamic radius of polymer molecules. The ratio of the intrinsic viscosity and the molecular weight provides information on the LLDPE branching level. The results obtained from FIPA showed a good correlation with the results obtained using the more laborious ASTM extraction method.

7.2.2 Control of Extractables in LLDPE Resins

Two approaches are used to produce LLDPE film with low hexane extractables suitable for food application [3, 8]. The first approach is to employ LLDPE resins produced with supported metallocene catalysts. These resins have both a narrow molecular weight distribution and a narrow composition distribution. As a result, the content of hexane extractables and xylene solubles in such resins is low.

An alternative approach is a chemical modification of Ziegler-Natta catalysts. The modification is usually carried out using selective catalyst poisons. Two types of the poisons can be employed. Poisons of the first type are electron donor compounds. They preferentially poison active centers I and thus reduce the content of the amorphous fraction in the resins. Although such poisons, as catalyst poisons in general, reduce the catalyst productivity, a careful control of the ratio between the amounts of the selective poison and the cocatalyst, triethylaluminum, allows one to achieve a desired level of xylene solubles or hexane extractables in the resins [4].

The second approach employs poisons that decrease the amount of active centers V in the catalysts [5]. These centers produce copolymer molecules with a very low α-olefin content (Table 7.1). When these active centers are eliminated from a catalyst, a lower concentration of an α-olefin is required to produce an LLDPE resin of a given composition and density. As a result, active centers I start producing copolymer material with a lower α-olefin content, a higher molecular weight, and lower solubility. Using such poisons afforded the synthesis of LLDPE resins with a greatly reduced content of the amorphous fraction of low molecular weight. In one example of using this approach, the content of hexane extractables was reduced from ~3 wt % in the resin produced with the original catalyst to ~0.5 wt % in the resin produced with the modified catalyst [5].

7.3 Blocking of LLDPE Film

Another technological property of LLDPE resins closely associated with their compositional nonuniformity is the phenomenon of film blocking. By definition, blocking is the tendency of two pieces of adhering polymer film to resist separation. Blocking is an undesirable feature for general-purpose LLDPE film.

Film blocking usually develops in two situations:

1. When blown or a cast LLDPE film is tightly wound into large rolls. Unwinding such rolls requires significant effort.
2. During storage of polyethylene packaging materials, when layers of flat smooth film (for example, polyethylene grocery sacks, trash bags, or merchandise bags) are in close contact.

7.3.1 Measurement Method

The blocking force is measured according to ASTM D 3354. It is expressed as a force (in grams) required for the separation of two pieces of film of a certain size. The test uses a special apparatus with two plates, one fixed and one movable. According to Procedure B (Standard Test Method for Blocking Load of Plastic Film by Parallel Plate Method), two pieces of film sized 4.3 by 4.3 in (~11 by 11 cm) are attached to the horizontally positioned plates and are brought into close contact.

The upper plate starts moving upward at a speed of 0.2″/min (5.1 mm/min) until the film pieces are separated by the 3/4″ (1.9 cm) distance. The force required for the film separation is measured. The upper limit of the test (the strongest measurable blocking) is 200 grams.

Empirical studies demonstrated that the blocking potential of LLDPE film depends primarily on the copolymer composition (resin density): the higher the α-olefin content in a resin (the lower its density) the higher is the resin's propensity to blocking [9]. Another factor affecting the blocking phenomenon is the melt elasticity of the resin, which is determined by the width of its molecular weight distribution (see Chapter 2). Resins with a lower melt elasticity (narrower molecular weight distribution) produce blown film with lower blocking potential.

Experimental evidence demonstrated that blocking force of LLDPE film is directly related to the content of xylene solubles or hexane extractables in the resins. In general, the higher the amount of xylene solubles or the hexane extractable content, the higher is the blocking force.

Several detailed investigations were carried out to determine the mechanism of LLDPE film blocking on the molecular level [10–12]. The principal conclusions from these studies is that the strength of film blocking is directly related to the compositional nonuniformity of the resins. Film produced from LDPE resins has a very low blocking force and film produced from metallocene LLDPE resins (materials with a high compositional uniformity) also shows low propensity to blocking. The highest level of blocking was observed in film from LLDPE resins produced with supported Ziegler-Natta catalysts. Atomic force microscopy studies linked the blocking phenomenon with the formation of a layer of soft amorphous material on the film surface [11]. This material migrates from the volume of the film to the surface during the film manufacturing and during the production or conditioning of various items made from film. The data in Table 7.1 show that this amorphous layer mostly consists of the copolymer component with a high α-olefin content, which is produced by active centers I in Ziegler-Natta catalysts.

7.3.2 Control of Film Blocking

Several techniques were developed to reduce blocking of LLDPE film. They encompass three different approaches:

1. Adding to an LLDPE resin a small quantity (10–15%) of a polymer material with a low blocking potential, such as an HDPE or a LDPE resin.
2. The use of antiblock additives during the film manufacture. The most common antiblock additives are silica or talc products with a small particle size, 3 to 5 microns. The amount of the additive depends on the thickness of the film and

the speed of its cooling; it usually varies from 0.1 to 1% with respect to the resin. The mechanism of the antilocking action of powdery additives is purely mechanical: their particles protrude from the film surface and prevent a close contact between surfaces of adhering film.

3. The use of slip additives, such as amides of fatty acids, for example, oleamide or erucamide. These additives are usually added to film resins in the amount from 0.1 to 0.2%. These organic substances have low solubility in amorphous components of LLDPE resins. When film is manufactured from such resins, slip additives rapidly migrate out of the resin and form a very thin double layer on top of the film surface, similarly to soap in water. The slip layer reduces the friction coefficient of the film thus allowing film surfaces to separate more readily.

7.4 Haze of Polyethylene Film

Haze is scattering of light passing through film or a sheet of an optically transparent material. When objects are viewed through a light-scattering material, a hazy (smoky) field around the light source is observed. Haze of polyethylene film is determined according to the ASTM test method D1003. The text describes details of instrumentation for measuring transmission haze, the procedures of haze measurement and the calibration methods. Two instruments can be used, a *hazemeter* (Procedure A, Section 5) or a spectrophotometer (Procedure B, Section 8). Equivalent European standards for the haze measurement are ISO 13468-1 and ISO/DIS 14782. Several modified procedures and calibration details are discussed in refs. [13, 14].

To measure haze of film, two parameters are compared, the total light flux passing through the film sample and the light flux deviating from the beam axis by less than 2.5°. Haze (in %) is calculated as the normalized difference between the two flux values. Different grades of polyethylene film can be roughly divided into two groups, film with low haze (between 3 and 10%) and film with high haze (from 10 to 20%).

The reason for difference in haze of polyethylene film grades is explained by their micro-morphological features described in Section 1.9 in Chapter 1. When an HDPE or a LLDPE resin with a low α-olefin content is crystallized from the melt, the solid material consists of tightly packed spherical bodies (spherulites) ranging in diameter from 10 to 50 microns – see Figure 1.17. The spherulites consist of smaller structural subunits, rod-like fibrils spreading in all directions from the spherulite center and filling its volume (Figure 1.16). Each fibril is the agglomerate of the smallest distinguishable morphological structures, lamellae, small flat crystallites formed by folded linear segments of polyethylene chains. The lamellae are interconnected by polymer chains that pass from one lamella to another (Figure 1.15 and Figure 1.16).

Each spherulite consists of three distinct phases, the crystalline phase (the core of lamellae), the completely amorphous phase (polymer chains filling the space between lamellae, see Figure 1.15), and a transitional semi-ordered layer between the crystalline phase and the amorphous phase. The presence of α-olefin units in LLDPE molecules inhibits uniform folding of polymer chains during crystallization because side groups of α-olefin units cannot be accommodated in the polyethylene crystalline lattice. The consequences of this hindrance are threefold (Figure 1.15): the presence of a small amount of α-olefin units in polymer molecules makes the lamella thinner (which makes the spherulites less rigid), it decreases the relative number of lamellae, and increases the volume of the intermediate and the amorphous phases. Table 7.3 shows the average thickness of lamellae, the thickness of their crystalline core, and the thickness of the intermediate phase in ethylene/1-hexene copolymers of different composition [15]. As the 1-hexene content in the copolymers increases, the average thickness of the lamellae and the thickness of their crystalline core rapidly decrease.

Table 7.3 Supermolecular Structural Elements of Ethylene/1-Hexene Copolymers Produced with Ziegler-Natta Catalyst [15]

$(C_M^{copol})_{av}$, mol %	Branching degree, CH_3/1000C	Lamellar thickness, nm	Crystalline core, nm	Intermediate layer, nm
0	0	21.6	14.3	0.6
0.7	3.3	16.5	9.1	1.0
1.0	4.6	16.0	8.9	1.0
1.5	7.3	15.2	8.2	1.1
5.2	22.7	9.9	5.3	1.2

Light dispersion (haze) in LLDPE film is determined by two factors:

1. The crystalline phase in lamellae and the amorphous phase in spherulites have a different refraction coefficient and disperse light differently.
2. This difference in the refraction coefficient is smaller when the lamellae are thin.

Both the overall thickness of the lamellae and the thickness of their crystalline core strongly depends on the compositional uniformity of ethylene/α-olefin copolymers. In the case of compositionally uniform LLDPE resins produced with metallocene catalysts, all the macromolecules crystallize equally poorly and form very thin lamellae. Film manufactured from such resins exhibits low haze. On the other hand, compositionally nonuniform LLDPE resins prepared with Ziegler-Natta catalysts contain a significant fraction of copolymer molecules with a very low α-olefin content, which are produced by active centers IV and V (Table 7.1 and Figure 7.1). These copolymer components form thick lamellae with well-organized crystalline cores. Such lamellae scatter light more strongly and, therefore, film produced from compositionally nonuniform LLDPE resins has significantly higher haze. As an example, film manufactured from an LLDPE resin (density 0.918 g/cm^3, I_2 melt index 0.9 g/10 min) produced with a commercial Ziegler-Natta catalyst has 10–18% haze whereas film made from a metallocene LLDPE resin with the same parameters has 5–7% haze [3].

References

1. Kissin, Y. V., *Macromol. Chem., Macromol. Symp.* (1995) 89, p. 113
2. Nowlin, T. E., Kissin, Y. V., Wagner, K. P., *J. Polym. Sci., Part A: Polym. Chem.* (1988) 26, p. 755
3. Cheruvu, S., Lo, F. Y., Ong, S. C., *U. S. Patent* 5 420 220 (1995)
4. Mavridis, H., Beran, D. L., Golden, J. R., Pater, J. T. M., Morini, G., *U. S. Patent Appl.* 2012 0053305 A1 (2012)
5. Ford, R. R., Stuart, Jr., R. K., *U. S. Patent Appl.* 2019 7652113 B2 (2010)
6. Kissin, Y. V., Brandolini, A. J., Garlick, J. L., *J. Polym. Sci., Part A: Polym. Chem.* (2008) 46, p. 5315
7. Van Rensburg, Q., Luruli, N., Sadiku, R., *Macromol. Symp.* (2012) 313, p. 43
8. Foster, G. N., Chen, T., Vogel, R. H., Wasserman, S. H., Lee, D.-C., Reichle, W. T., Karol, F. J., Whiteker, G. T., *U. S. Patent* 6 159 617 A (2000)
9. Hazlitt, L.G., Karande, S. V., Castille, M. J., *J. Appl. Polym. Sci.* (1994) 51, p. 271
10. Vincent, O., Osmont, E., Sommer, F., *Society Plast. Eng. Meeting, ANTEC 2002,* San Francisco, CA, 2002
11. Vincent, O., Osmont, E., Sommer, F., *J. Plast. Film Sheet.* (2004) 20, p. 247
12. Najarzadeh, Z., Ajji, A., *J. Plast. Film Sheet.* (2017) 33, p. 235
13. Yu, H.-L., *Meas. Sci. Technol.* (2006) 17, p. 29
14. Yu, H.-L., Hsaio, C.-C., *Metrologia* (2009) 46, p. S233
15. Marigo, A., Cingano, G., Marega, C., Zannetti, R., Ferrara, G., Paganetto, G., *Makromol. Chem. Phys.* (1995) 196, p. 2537

Index

F

G

H

I

L

M

N

O

P